The Indispensable Guide to Undergraduate Research

The Indispensable Guide to Undergraduate Research

Success in and Beyond College

Anne H. Charity Hudley
Cheryl L. Dickter
Hannah A. Franz

Published by Teachers College Press, 1234 Amsterdam Avenue, New York, NY 10027

Cover photos (left to right): SolStock; monkeybusinessimages; Steve Debenport; and m-image photography. All are via iStock by Getty Images.

Library of Congress Cataloging-in-Publication Data is available at loc.gov

ISBN 978-0-8077-5850-2 (paper)
ISBN 978-0-8077-7582-0 (ebook)

Printed on acid-free paper
Manufactured in the United States of America

24 23 22 21 20 19 18 17 8 7 6 5 4 3 2 1

Contents

Acknowledgments

Prof. Charity Hudley writes this book in honor of Drs. Cynthia and Renard Charity and in memory of Mr. and Mrs. Alfred and Sarah Charity and Mr. and Mrs. Leslie and Annie McClennon. Special thanks to Mr. J. Chris Hudley, Mr. Renard Charity Jr., and Mrs. and Mr. Renée and Mike Price. This book is truly for my nieces and nephews Madeleine, Emma, and Olivia Charity and Carter, and Caroline Price. I hope it sees you well through your high school and college years. Thanks also to the Hudley family, particularly Mrs. and Mr. Marie and Jay Hudley and Mrs. and Mr. Tiffaney and Jeremy Armstrong for their constant and loving support.

Prof. Dickter writes this book in memory of her father, Bruce Dickter, and for future Class of 2036 Scholar Greyson Greene. She thanks Dr. Kristen Klaaren for her endless encouragement and Mr. Gerris Greene, Mrs. Leah Greene, and Mrs. Marie Dickter for their amazing love and support.

Ms. Franz writes this book in honor of her former students at Lake Taylor Middle School and the teachers there who mentored her, especially Mrs. Letitia Frank, Mr. Larry Carr, and Mrs. Carolyn McCoy. Thank you, Matt Franz, for all your feedback and encouragement. Thanks to Stella, Peter, Eliza, and Mark Askin for your love and support. A very special thanks to Hank Franz for keeping me company while I wrote this book and for being my motivation to continue working for all students.

We owe immense debts of gratitude to our mentors and teachers: Dr. Lee Perkins, Dr. William Labov, Dr. Calvert Watkins, Dr. Hollis Scarborough, Dr. Gillian Sankoff, Ms. Patty Cassella, Dr. Bruce Bartholow, Dr. Walt Wolfram, and Dr. Leslie Grant. Without each of you, we would not be the scholars we are today.

We are indebted to our editor Mr. Brian Ellerbeck, to Ms. Jamie Rasmussen for publicity and marketing assistance, Ms. Lori Tate for production assistance, and Ms. Christine Crocamo and Mr. Karl Nyberg for copyediting assistance. We are further indebted to our manuscript reviewers, who gave us excellent feedback that improved this book. We also acknowledge Christine Fulgham, who took our photo for the book's online resources.

We offer heartfelt thanks to each of the scholars who contributed vignettes to this book: Dr. Rhonda Fitzgerald, Ms. Lauren Collier,

Ms. Mikaela Spruill, Mr. Nathanael Paige, Ms. Ebi Doubeni, Ms. Rachel Brooks, Mr. Brian Anyakoha, Mr. Dondré Marable, Ms. Danielle Weber, Ms. Katie Mitchell, Ms. Thomeka Watkins, Ms. Christine Fulgham, Ms. Kristin Hopkins, Ms. Sara Taylor, Ms. DaVon Maddox, Mr. John Nguyen, Ms. Ebony Lambert, Mr. Marvin Shelton, and Mr. Amirio Freeman. We also thank our wonderful colleagues and students who read drafts of this book and shared invaluable advice and suggestions: Ms. Destini Brodi, Ms. Jasmine Brown, Ms. Ebi Doubeni, Mr. Amirio Freeman, Ms. Ebony Lambert, Ms. Heather Lawrence, Dr. Christine Mallinson, Mr. Nathanael Paige, Mr. Cleveland Winfield, and Dr. Sharon Zuber. We are so grateful to the students who participated in our research groups and took our research-based courses.

We especially thank the scholars who attended the first 6 years of the William & Mary Scholars Undergraduate Research Experience (WMSURE), particularly those who served as formal and informal fellows and mentors to other scholars, including: Mr. Jerome Carter, Ms. Ebony Lambert, Ms. Vanessa Gray, Mr. Marvin Shelton, M. Edward Hernández, Ms. Eboni Brown, Mr. Dondré Marable, Mr. Adom Whitaker, Ms. Dahanah Josias Sejour, Ms. Ebi Doubeni, Mr. Jacob Abrams, Ms. Darlene Dockery, Ms. Melanie Lichtenstein, Ms. Jasmine Koech, and Ms. JoEllen Blass. We learned so much from hearing what each of you had to say. We also thank our WMSURE faculty mentors, who were instrumental in planning for this book, including: Ms. Artisia Green, Dr. Monica Griffin, Ms. Susan Grover, Dr. Chon Glover, Dr. Vernon Hurte, Dr. Chris Howard, Dr. Elizabeth Harbron, Ms. Laura Heymann, Dr. Shantá Hinton, Dr. Sharon Zuber, and Ms. Natasha McFarland.

This text would not have been possible without the inspirational models of the University of Pennsylvania Africana Studies Pre-Freshman Program. We are grateful to Dr. Herman Beavers and Dr. Tukufu Zuberi, the Thurgood Marshall Fellowship at Dartmouth College, with particular thanks to Dr. Ioana Chitoran, and the McNair Scholars Program, particularly at the University of Pennsylvania, with particular thanks to Dr. Malcolm Bonner.

We appreciate the feedback, assistance, and encouragement that we received from other scholars and friends, many of whom read and listened to sections of this book manuscript, including Dr. Mary Bucholtz, Ms. Kirsten Bradley, and Ms. Catherine Hartman.

We recognize our colleagues at the College of William & Mary, particularly our deans and department chairs, who were instrumental in providing us with support to complete this project, particularly William and Mary's Provost Michael Halleran, Dean Joel Schwartz, Ms. Jabria Craft, Ms. Diana Healy, Dean Virginia Torczon, and William and Mary's former Dean Carl Strikwerda, who is now president of Elizabethtown College. Prof. Charity

Hudley also recognizes her colleagues at the University of California Santa Barbara, particularly Chancellor Henry Yang, Executive Vice Chancellor David Marshall, Dean John Majewski, Vice Chancellor Claudine Michel, Dr. Mary Bucholtz, Dr. Matt Gordon, and UCSB's former Dean Melvin Oliver, who is now president of Pitzer College. We also thank former colleagues, professors, and mentors at Dartmouth University, Harvard University, North Carolina State University, the University of North Carolina at Chapel Hill, Randolph Macon College, Union College, and the University of Pennsylvania.

Our current and former students have been a great resource. We particularly acknowledge our many undergraduate research students.

We have presented material from this book at various conferences, including the Linguistic Society of America Conferences (2013, 2016); the Literacy Research Association Annual Conference (2016); the Linguistics Society of America Summer Institute (2015); the National Council of Teachers of English Annual Convention (2014); the American Association of Colleges and Universities Conference (2012); the College of William & Mary Workshop on Tenure (2011); the American Dialect Society Conference (2010); and the National Science Foundation Workshop on Broadening Participation (2008).

We also benefited from scholars and students who gave us their feedback on material from this book at talks given and posters presented at the Council of Undergraduate Research Biennial Conference (2016) and the National Conference on Students in Transition (2016).

Special thanks to all of the faculty and staff who have attended our duPont Fund–sponsored WMSURE faculty conferences, and to our presenters: Mr. Michael Mallory, Dr. Joy Davis, Dr. Joseph Williams, Dr. Keivan Stassun, Dr. Benjamin Castleman, Mr. Tyren Frazier, Ms. Charleita M. Richardson, and Mr. Eder Rivas-Hernandez.

Prof. Charity Hudley acknowledges Brown University; Emory University; Georgetown University; The Ohio State University; Johns Hopkins University; Kansas State University; Norfolk State University; North Carolina State University; Northern Virginia Community College; The University of California at Santa Barbara; University of Michigan at Ann Arbor; Richard Bland College of William and Mary; University of Virginia; Virginia Commonwealth University; and Virginia State University for their support of this work.

Parts of this book draw upon or are adapted from material we have published elsewhere. We acknowledge Prof. Charity Hudley's editorship of Language Teaching Linguistics and her following chapters: "Linguistics and the Broader University" (Charity Hudley & Mallinson, in preparation), "Linguistics and Social Activism" (Charity Hudley, 2012), "Linguists as

Agents for Social Change" (Charity Hudley, 2008), and "Service Learning as an Introduction to Sociolinguistics and Linguistic Equality" (Charity Hudley, Harris, Hayes, Ikeler, and Squires, 2008).

Our work has been supported by various funding sources. Prof. Charity Hudley acknowledges the State Council of Higher Education in Virginia (SCHEV) Capstone English Project, Senior English Academy, and Visible Leaders grants; the QEP Mellon Initiative at the College of William & Mary; the College of William & Mary Community Studies Professorship, and the Class of 1952 Professorships. Prof. Charity Hudley's research has also been supported in part by the National Science Foundation grants BCS-105105 and SES-0930522. Prof. Dickter's research has been supported by the National Institutes of Health, the National Science Foundation, the National Psi Chi Honor Society, William and Mary Summer Research Grants, the Eastern Virginia Medical School/W&M Collaborative Grant Program, the Creative Adaptation Fund, and the Suzann Wilson Matthews Research Award. She also acknowledges the support the College of William and Mary has provided through the Wilson P. and Martha Claiborne Stephens Term Distinguished Associate Professorship. We also thank funders to WMSURE: the Jessie Ball duPont Charitable Foundation, the Bank of America Charitable Foundation Grant, the College of William and Mary, and many generous donors.

Our deepest thanks go to the individuals, organizations, and institutions we have mentioned and any we have inadvertently omitted. We are truly grateful for your support.

Preface

Setting the Stage for Success

WHAT IS RESEARCH?

For the purposes of this book, we are defining research as an inquiry or investigation that makes an original, intellectual, or creative contribution to a discipline, area, question, challenge, or theme. Let's unpack this definition a little. The first part is the inquiry or investigation, which refers to the pursuit of an answer to a research question, or what you are trying to find out or discover by conducting your research. The second part is the contribution that you can make by new, innovative, or creative thinking or processing. As you will see, this contribution can take many different forms and will differ based on discipline. In general, research projects should not be identical to research that other people have done: Your research should involve the creation and discovery of new knowledge. Most undergraduate research should have the goal of sharing these new findings through presentations and performances or in publications.

This book involves some metacognition. *Meta* comes from the Greek prefix meaning "after" or "beyond" and is often used to mean "another layer of." So in combination with *cognition*, *metacognition* means "thinking about your thinking." To be metacognitive in your learning is to be introspective and reflective about your learning process. Not only do we ask you to think about how you think and learn, but we also discuss what we might call meta-research, or research about research, throughout the book. Thus, the research we discuss both supports our ideas for you about your own undergraduate research path and provides you with different examples of how research can be done. Throughout this book, we invite you to be especially meta and to reflect on how the information we provide in these pages pertains to you personally. We also encourage you to talk about the material with others and to ask for others' opinions about these topics.

WHY DO RESEARCH?

This book is focused on research because it is how we best learn and because it is the best that colleges and universities have to offer. Research is how innovators both at universities and in other places create and discover knowledge. Research is the foundation of the modern university because the process of discovery is important not only to those who study a particular topic but for the broader world (Booth, Colomb, & Williams, 2008). Research reveals what we don't yet know and helps us to move forward in our social and physical world. This book is designed to help you understand why getting involved in research can be such a crucial part of the college experience. If you are reading this book you are already brave and inquisitive. You or someone who cares about your education has suggested that you learn more about the undergraduate research process. You can do research as part of your classes, outside of the classroom throughout the academic year, and at your school or at another location in the summer. Research can give you academic credit toward graduation or you can get paid for it. Throughout this book, we share information with you on why and how to get started on research while you are an undergraduate student. Our students (designated as scholars throughout the book) have helped us prepare this book and, throughout each chapter, you will see quotes from them about how they have learned to best navigate the college experience and why research has proven critical to their success. As you transition from high school to college, you will experience changes in scheduling, class structure, and relationships with teachers and peers. So many changes all at once often lead to some degree of ambiguity about how to navigate your new environment. It may be unclear whom to ask for assistance in managing such experiences, and we can help you with that. It is important to learn about the questions that *you should be asking* as you make this transition. This book is designed to help you identify what questions to ask—and whom to ask—about scholarly life and the transition to college with an emphasis on research as a tool for success. Asking the right questions can lead you to discover answers about your own unique role at your college or university, which is often referred to as the "academy," and how research can fit into your success in the academy.

As scholar Lauren Collier explains:

> Participating in research as an undergraduate has benefited my life in several ways. In the short-term, it allowed me to develop both academically and professionally while I was still in school. Skills learned in the lab such as written and verbal communication, time management, team building, and organization have proven to be instrumental to success in

my college classes and work life. In the long-term, conducting research allowed me to clarify my career and educational goals.

One exciting aspect about research and about starting your college career is that the more questions you answer, the more questions you may have. Such an experience may seem counterintuitive but reflects critical thinking as well as academic and personal growth. College is a time to ask questions and to get the answers to questions you have always had. This process can happen through your classes, but you can take critical thinking to the next level in out-of-class research experiences. Research means discovery, and integrating discovery into the college experience moves you from being a consumer of knowledge to a creator of knowledge. To that end, this book is designed to help you be a research scholar rather than just a college student.

Bain (2012), who writes about "what the best college students do," emphasizes the creativity of what he deems the "best students" find during their college years. Bain does not equate the creative experience with academic or financial success, although this success is important and we will talk about ways to help you achieve this success throughout the book. Rather, Bain's focus is on the creative and transformational experience that can happen when new knowledge and new experiences result in new ways of thinking about and seeing the world.

The research experience, more than any other academic endeavor, illustrates that there is no one-size-fits-all approach to succeeding at college or being successful after college. Our students have found all different ways to succeed, and part of the challenge of college is trying out different strategies for success and identifying what works for you. Throughout this book, our students will share what worked for them (and what didn't), and we will give you advice on how to figure this process out sooner rather than later. It's important to know that what works for you might be different from what works for someone else at your school. You will eventually figure out how to excel at school, but realize that it may take you a little time. Success in college and after includes learning how to deal with failure and how to reach out to others for support and mentoring (Bain, 2012; Pollak, 2012). This book will provide you with resources for doing just that and will help you incorporate research into your success strategy. Success is often preceded by failure, even if it's a small failure, and overcoming failures can be crucial to your college success and your personal growth.

As scholar Mikaela Spruill says:

As the epitome of a well-rounded student, I came into college with nothing but wins under my belt. From athletics to academics, everything I

> attempted came easily to me. I was accustomed to excelling with little effort. Then, I stepped into the whole new world of research and after picking up on a few things, I figured I had caught the gist of this whole research thing. So with an interesting observation in my pocket and a brain full of ambition, I presented my hypothesis to my brilliant advisor and became an independent researcher working on my own project. I had no clue what I had gotten myself into. My first study revealed that my data did not support my hypothesis, and this rocked the world of the kid who had skated through everything with ease. For the first time I had to truly decide whether to push through or give up. Although failing was disappointing, it ignited a fire in me and I was determined to succeed the next time. The research process helped to mold me by teaching me the definition of perseverance. By overcoming the obstacles placed before me in the lab, I learned to rise above other challenges on campus and in my life. Undergraduate research taught me how to graciously fail and, more so, how to resiliently triumph.

Whether you have experienced some sort of failure before or have never failed at anything, developing skills for coping with situations in which you do not succeed as much as you would have liked is part of the research process. This book will describe how you can use research to help you learn where to find the best supports and opportunities available at your institution and at institutions throughout the world.

Over the years, most universities have relied on professors and graduate students to conduct the majority of research at the college level. Research at your school is conducted by professors, most of whom have completed years of research as part of the requirements for their masters and/or PhD degrees. To make universities even greater, more recently colleges are now putting a greater emphasis on undergraduate research, which means that students like you have increasing opportunities to get involved with research, to learn how to create and discover for yourselves, and to add new knowledge to a field. Inquiry and discovery are so important at colleges, and institutions are recognizing that you as an undergraduate student can offer a unique perspective and contribute to the research process at many different levels.

Throughout this book, we will share with you our research process in writing the book. We will have textboxes with questions for you to answer, reflect on, and write about. We'll also include sources for the material in our text that will be linked on our website so that you can find out more about the information and its authors easily. We have a full bibliography at the end of the book. That is, while conceptualizing and writing this work, we relied on research conducted by others and ourselves so that we could demonstrate a research-based approach to getting started with undergraduate research. Basically, we did research about research! We review evidence demonstrating the benefits of research and how to use research to be successful in college, and we provide examples from faculty and students about their own experiences. Our goal in writing this book is to share both the academic and social aspects of research, which is an academic process and a social journey.

We describe research on the journey from high school to college from different perspectives (ours, our students', and yours) with a mind toward graduating from college with academic and research honors (we explain the research honors model in Chapter 2). We outline several of the important differences between learning and succeeding in high school and learning and succeeding in college, based on our research and the research of others. To illustrate the differences, we provide descriptions and tips from current and former college students who have been through this transition as well as from professors who have mentored students.

We will also identify and describe more specific challenges for students who are members of social, racial, and ethnic groups that are underrepresented in higher education, including African American, Latinx, and first-generation students (see Textbox 1.2 in Chapter 1 for working definitions of these categories). Our research has revealed that students from underrepresented backgrounds are less likely to participate in research experiences in college than students from overrepresented or represented backgrounds. We think this difference is largely because fewer engaging research opportunities are presented to students from underrepresented backgrounds. We'll explain why and what you can do about it. We are here to help you disrupt that narrative and change your future and the future of other scholars forever. This part of the book is important regardless of whether or not you are a member of the underrepresented groups we focus on, because being aware of these issues will help you make a way for your peers, those who come after you, and those ahead of you with whom you may work.

The Indispensable Guide to Undergraduate Research

CHAPTER 1

Finding the Tools to Become a Scholar

In this book we take a critical-thinking, research-based approach, rather than a how-to approach, to guiding you through the undergraduate research process. Our goal is not to tell you what the answers are so that you might become an undergraduate researcher; it's to help you figure out what questions to ask. One of the reasons for our approach is that there is no one model of research success that will work for everyone, and there are different types of information that will benefit some more than others. Another reason is that, this way, you are part of the critical-thinking process. We want you to be able to clearly and purposely decide what questions to ask to meet your needs. This critical-thinking model parallels the research process, which is similarly fluid and infinite. Often the goal of research is not a definite answer, but a better set of questions or a creative exploration. Through understanding how to ask questions, you understand the scholarly process and become part of the scholarly community; you learn how people ask questions and what research entails. Asking questions is not only important for helping you transition to college and academic research, but is also crucial in helping with the college-to-career transition. (See Pollak [2012] for more about asking questions as you move into careers.)

As scholar Nathanael Paige notes:

> Research is a journey—the more questions you answer, the more questions you have. Finding something new can lead to more discoveries. Keeping an open mind during the research process is important to get the most out of this journey.

The Indispensable Guide to Undergraduate Research is designed especially with undergraduate research in mind to help you figure out what questions to ask along the way as you do research in college. We want to make sure that you know about the opportunities that are available to you. What does it mean to be a research scholar, and why would you want to be one? What's in it for you? We will discuss the many benefits of conducting undergraduate research; furthermore, we will cite research conducted at a variety of universities to provide scientific evidence demonstrating these benefits.

One good place to start working on becoming an undergraduate research scholar is to ask and answer the following questions:

How many more lives can you touch as a college-educated researcher? (including your own!) You can do so much with a college education. You can help others. You can earn or make money. You can teach and mentor others, both formally and informally. You'll be able to help others both appreciate and question the world around them. You can lead. In so many ways, the students that come after you will also benefit from your experiences and the direct and indirect impact that you have on your college. You can come back as an alum (a graduate of your college) and inspire the next generation of students.

What jobs can you do that require a college education and research experience? We encourage the doctor, lawyer, businessperson professional perspective on what career opportunities college can open up, but we also want to encourage so much more. We want to see you become the future's college leaders, researchers, educators, and policymakers. We encourage you to create a job that fills a need you find in your own research. And even in the traditional jobs, research is necessary for you to get in and get the job and then be successful at it. Many of the skills that employers look for overlap with research skills, including independent thinking, taking risks, and working toward success despite ambiguity (Kinkel & Henke, 2006).

What insights can you gain as a college-educated researcher? As you will learn in this chapter, as a college-educated researcher, you will be open to more insights because of your skills and experience. You will be able to read and summarize and critique. You will have the insight to avoid re-inventing the wheel through both the knowledge you have as a student *and* the knowledge you create as a scholar. Through these insights, you will be able to improve the quality of life for you and others around you.

What things can you change, not only because of your intellect, but also because certain types of jobs require that you have research experience? Social change, economic change, all kinds of change for the good (and notably also for the bad) have been achieved by scholars. You will gain an understanding of how research becomes policy in both the private and public sectors. For example, as described in more detail in Chapter 2, finding the answers to questions that are important to you personally and that can change our society and the world can be extremely rewarding. You can use your research as a tool to talk to others about the importance of different issues and use what you've learned to implement change.

What will be different about your own life when you are a researcher? With the insights you gain from your research and the research process, what will you question rather than just believe or take for granted? How will you see yourself as part of the process? Research can make you a critical thinker and help you examine the world around you with a more critical eye. For example, when you hear a statistic reported in an online news article, you will be better able to evaluate its validity.

How much more good can you do for the world and what can you contribute both to your local community and the world at large? It is important to remember that "good" can be defined in a myriad of ways. Good is intellectual, financial, social, and spiritual, to name just a few perspectives. We wait anxiously for the combination that you will create! We'll ask you now and all through this book: What do you want to be when you grow up? It was an important question in kindergarten and it's an important question now, as college expands the range of possibilities of both who and what you can be.

What ideas can you create, models can you build, and solutions can you discover that have never been thought of before? Learning the content of courses is just the beginning; synthesizing it and building upon it is the name of the game. We write this book to encourage you to be masters of knowledge and challengers of the status quo. During your college years, be devoted to being both thinkers and dreamers! College is a time in your life when you can try most anything—so do! Research opportunities that are available at colleges and universities will help you be prepared to try anything. Research is all about creativity and innovation—it takes you from being a knowledge receiver to being a knowledge creator.

Many books and articles that focus on college success aim to help prepare students to obtain good grades, overcome obstacles, and graduate in good standing prepared for a career. The focus of this book, however, is to help you *excel* in the university setting. For many of you, it is likely that you will work hard and succeed so that you achieve strong grades and learn much from your courses and activities. We want to stress in this book, however, that although it is important to be a *successful* college student, we think that it is extremely important to have the goal to *excel* in college and to reach your highest academic potential. We think of this difference as the difference between being a student, in which you learn important information and graduate with a degree that should help prepare you for the world, and being a scholar, in which you reach your greatest potential and are well prepared for the best graduate schools, professional schools, and workplaces. Part of what we think will make the difference between being a student and being a scholar is participating in research experiences.

STUDENTS VERSUS SCHOLARS: WHY COLLEGE IS NOT HIGH SCHOOL

Textbox 1.1 describes a student versus a scholar. As scholar Ebi Doubeni says:

> Coming to college, I just assumed that since I had good grades in high school it would be just as easy to get good grades in college. I soon found out that this was not the case. I just figured that I needed to study harder in order to get better grades but I didn't realize that I needed to change how I studied. Instead of attempting to memorize everything just to spit it back out for an exam, I started to work on actually learning the information by doing a closer reading of class materials and looking up additional information on the topic outside of class. I learned that to truly understand something you have to go above and beyond what was assigned in class and do an in-depth analysis of the subject. I learned that everything that I tackled in college should be treated like research, so now I don't distinguish my own personal research from my assignments in class.

Although college courses best accommodate students, many parts of college, especially research opportunities, are designed to help you develop as a scholar as well. This book will help you learn how to take advantage of those opportunities. To make it possible for you to develop as a scholar, it is important to think about some of the possible differences between your high school and other previous educational experiences and the experiences that you will have during college. Being a student relies on the work of scholars, but often in K–12 education and even in college the material may be dissociated from the scholar so that pathway is more obscured. For example, it is important to ask: Who wrote the material in your textbooks? What evidence do you have that the material is true? How do you know? These types of questions are part of what we call critical thinking—part of being a student but crucial to being a scholar.

A new college student, Lucian (a pseudonym), explains how he began to learn the difference between being a student and being a scholar as he moved from high school to college:

In high school, the readings they gave you were not necessarily to develop your idea on the subject as a whole. It's more, "Here's the material that we're going to be testing you on." I think some kids struggle transferring from high school to college because they may be preparing for the test instead of actually learning about the subject.

Textbox 1.1: Student versus Scholar

The nature of a student's work:

- A student learns what is taught. Instructors' requirements drive what you have to know.
- A student focuses on mastery of class content. Information that is presented is broken up into distinct classes, and while a major may be cohesive, individual classes don't have to relate to each other at the time. Faculty may have a sense of what should be required but not work comprehensively together on content integration.
- Being a student is even important to being research-ready because you have to know something to get ready to have an independent focus: how to write, compute, design, perform, etc.—so being a student is not a devalued thing, we just don't want you to think of it as the end of the process. It's actually just the beginning.

The nature of a scholar's work:

- A scholar engages in research. Research can be in classes or outside of classes or even outside of your university.
- A scholar has independent focus and wants to learn outside of what a particular course or major demands. Research connects classes and at the highest level establishes students as authorities in areas where their professors may not be.
- A scholar asks lots of questions! Questions that elicit multiple answers, questions that elicit more questions, and questions that have not been asked before.
- A scholar makes connections between different classes or disciplines and makes connections between research and what is learned in the curriculum.
- A scholar authors their own work and process—not just the learning process, but the actual material that students learn.

Research will help you be able to make this change in mindsets in your transition to college. Academic culture is constantly shifting and changing, and that is the nature of being at a university. As Lucian touches on, in high school, school or state standards determined much of what you had to know, and most students studied according to a predetermined program and schedule. In college, in many ways, you ultimately decide what you need to know and how and when you will learn it. It can be challenging for new college students to learn that there is no right or wrong way to undertake their learning and scholarship process, but it can also be exciting to study topics that you didn't know were possible for academic work. You can study what you are interested in, and the possibilities for what you can study increase if you engage in research. If you learn how to take advantage of and responsibility for these opportunities early in your college career, you will be well prepared for life after college as well.

Doing research helps you learn how you best ask questions and how you can be an active part of your learning experience. For example, psychology professor Mary Dosch and education professor Margaret Zidon wanted to learn how well college students perform when the professor uses a one-size-fits-all lecture approach compared to a differentiated, or more personalized, approach (Dosch & Zidon, 2014). A differentiated approach takes into consideration the background knowledge, past experiences, and skill level of students, which can help students learn new material. Students can explore topics of their own choosing, which can fit nicely with their past experiences and interests. The authors conducted a research experiment and their results showed that students in the differentiated class did better on assignments and exams than students in the one-size-fits-all class. Regardless of the kinds of classes you take in college, you can make sure your learning experience is personalized to your interests, goals, and skills by engaging in research. As scholar Rachel Brooks notes, her research experience was extremely personal:

> One of the biggest differences I have noticed between my high school and college research is the opportunity—and even encouragement—to make work more personal. I remember being told to avoid "I" at all costs in a high school English class, but my coursework and research in college deconstructed that idea and declared that it is good to be personal and vulnerable in your work. For example, studying college readiness as a first generation college student offers insight into the struggles that others may lack. In my experience, making research less generalized and more personal clarifies its purpose and maximizes its value.

FACULTY-STUDENT VERSUS TEACHER-STUDENT RELATIONSHIPS

In college, individual professors determine much of what you need to know for their own classes and research groups. Being successful in their classes often takes a lot more than attending classes and doing well on exams. Each professor is different and each will have different expectations of students in their courses. To begin navigating the unknown, start being a scholar on day one by asking your professors questions so that you understand their expectations. That means asking about things big and small as well as small things that may end up big. For example, start by asking what the professor wishes to be called both in person and in email. Some prefer "Professor," some prefer "Doctor" if they hold a doctorate (ask if they do), some prefer "Ms." to "Mrs." or "Miss," others prefer to be called by their first name (again, ask!). We've seen some professors stiffen when the title they've worked years to earn (Dr. Smith) is replaced by the more common title (Mr. Smith), while others feel that fancy titles obscure the relationship between professor and student. It is also important to ask about their policies and philosophies on the use of laptops, cell phones, and tablets. Professors have different rules, and it's important to understand the rationale behind why you should or shouldn't be online during a lecture and/or discussion. If you take notes on your smartphone or tablet, some professors may not mind, while to others it might appear as if you are not engaged and are texting or on social media. Other things may also change from class to class. For example, in completing assignments, some professors may allow you to use sources to help such as the textbook or the Internet but others may want you to work on your own. There can also be ambiguity in general policies. For example, if plagiarism has been explained to you but you are still unsure of all that it entails, ask! If it hasn't been explained to you yet, ask! You will take these habits of questioning with you throughout your time in college and your future careers. If you don't want to ask these questions during class, meeting with your professor outside of class is a great option. We'll talk about that option more in Chapter 4, which focuses on the relationships that students have with their professors.

As a scholar, you can benefit in many ways from maximizing your contact with professors both inside and outside of the classroom so that you can best understand what is expected of you. Many professors, in addition to teaching classes, spend a lot of their time directing research groups or labs. Our strongest students are present in both contexts. Ask the professors who teach the courses you are most interested in if they do research with students and how you can get involved. It is important to ask these questions very early—in your first semester—because some professors will require you to take several courses before you can participate in research and you will need to plan ahead. Other professors may have research opportunities for you to get involved in right away, and these opportunities can help you learn how to be a scholar and reach your full potential in college from the beginning. See Chapter 2 for more specific advice on how to pursue research experiences with professors at your university.

With such opportunity also comes the responsibility of making sure your requirements are met within your classes, your major, and your college's graduation guidelines. It is up to you to ensure that your readings and practice problems and exercises are completed before class; you have to make sure that you understand the syllabus and the information provided. It is up to you to make sure you have all that you need to be successful and that we as faculty understand explicitly what your individual definition of success is and that you understand our individual definitions as well. So ask! We'll also put a plug in here for taking a less traditional path. That is, many schools allow you to craft your own major based on your personal interests and career goals (i.e., self-designed majors), and you can also focus on research and learning in the context of helping your community (i.e., service learning and community engagement).

COPING WITH SUCCESSES AND CHALLENGES

Another crucial aspect of being a scholar is understanding how to cope with both successes and challenges. In high school, you were near the top of the class as defined by your school. That's part of how admissions works for colleges. Now that you're in college, everyone was near the top of their high school class; now the definitions of top will differ dynamically. There is a deep pool of diverse ability among you! People can do what you do, and some can do it better. But you can do some things better than anybody! You may be able to do stuff that we haven't even thought of yet!

For many of you, this change will mean a change in your interpretation of success. This change will also mean a difference in your interpretation of study skills. For those of you who didn't have to take notes, create study guides, or even do the readings in high school, you will probably have to do so now to remain at the top of the class. Learning how to synthesize large amounts of information and how to create unique and strong arguments are critical to the scholarly process. Professors are more than willing to talk with you about the best ways to prepare and study for their classes and to refer you to students who have taken their classes for peer advice. Xavier University in New Orleans, LA, has a great way of thinking about the diversity of academic experiences that you may have had prior to coming to college:

> What doesn't work is saying, "You need remedial work." What does work is saying, "You may be somewhat behind at this time but you're a talented person. We're going to help you advance at an accelerated rate." (Steele, 1992)

We too know that you are talented, so if you find that you need to advance at an accelerated rate, please let your college know. Many offices on college campuses—Student Support Services, Learning Centers, Writing Centers, Deans of Students Offices—are there to support that transition and process. You should be using them ALL as resources regularly, not just when challenges arise. For example, you don't want your first visit to your professor's office hours to be after you have had challenges with an assignment or exam. Professors really appreciate your taking the initiative to talk to them about their class and especially your asking follow-up questions about the material. After all, we are interested in the topics we teach—this is why we teach these topics in our classes! Remember also that one of the reasons that professors chose to go into their profession was to work with students. We enjoy meeting and getting to know our students. Positive interactions between students and professors can be enjoyable for both parties. Get to know your professors at the beginning of the semester so that the sky remains the limit. Scholar Nathanael Paige talks about how he used on-campus resources to ensure academic success:

> During my first year of college, building a network of resources was a big success for me. I utilized that first year to acquaint myself with not only other freshmen, but also upperclassmen and faculty members. My biggest first-year challenge was directional focus. There are so many academic options and extra-curricular paths you can take in college, and it is easy to feel like you need to choose a path immediately, because everyone else seems to have confidently found their niche. Fortunately, if you patiently allow your passions to narrow down your search and use your professors and students as resources, everything will fall into the right place.

WHO WE ARE, WHO YOU ARE

Because of its intricate nature, the scholarly process is built around relationships between professors and students, so we'd like to start off by introducing ourselves and sharing with you why we wrote this book. These are the stories behind why we are so dedicated to the work that we do and the topics that we research.

Anne H. Charity Hudley

I grew up in Varina, VA, a rural area zoned for agriculture just east of Richmond. My grandmother, who was also from Varina and lived her adult life in Charles City, VA, thought that the College of William & Mary should educate more African American students—especially the ones who lived near the college. She didn't live to see that dream to fruition, but I have worked to see her dream through. To maintain the integrity of her vision, my goal is to be successful as a scholar and as a faculty member in a way that is universal yet respects my grandmother's vision of caring for communities. Such work honors the integrity and dreams of others like my grandmother who were never afforded a position to become scholars and researchers.

The College of William & Mary is just 45 minutes down the road from Varina, where my parents still live. I am from a historically multiracial background (African American, Native American, and White), but in many senses, in Virginia, the one-drop rule still persists, and I am proud to be African American. In this book and in the academic narrative, I represent the prep-school-to-professor experience. I attended St. Catherine's School in Richmond, VA for 13 years and I had an early interest in studying linguistics and in being a college professor and administrator. I was granted early admission to Harvard and found myself surrounded by supportive faculty and students. In many ways, this book is a product of my Harvard experience and the time over 3 years that my undergraduate professor Calvert Watkins took in helping me write my honors thesis and develop as a scholar. My thesis explored the idiolect of Bessie Smith—the ways in which her individual language and singing style changed over time. I gained a great introduction to African American language and culture in the South, and it became the very important start to the work that I'm doing today.

Yet, at Harvard, no one in my department told me about the Mellon-Mays Undergraduate Fellowship Program until it was too late for me to apply. From then on, I found many of the fellowships and grants that I earned on my own, and that process is described in this text. My family has always been very important to me, and my grandparents were strong role models. I am an example of what can be accomplished in the third generation of African American scholars. My father taught at Columbia and my mother did research with a Nobel Prize–winning professor as a first-year student at Vanderbilt. My brother, sister, and I have all worked with leaders in our respective fields: My advisor was Calvert Watkins at Harvard University, my brother's was Nell Painter at Princeton, and my sister's was Deborah McDowell at the University of Virginia. My research on the language and culture of high-achieving underrepresented students forms the underpinnings of this book.

At the University of Pennsylvania, I began studying in earnest how discrimination based on language and culture led to educational inequalities. I also became very interested in the transition from high school to college and undergraduate research through my work with The Center for Africana Studies Summer Institute for Pre-Freshmen and the Penn McNair Scholars Program. This book got started at the University of Pennsylvania in these summer programs. At the time, I thought my interests in linguistics and in supporting underrepresented students in undergraduate research were somewhat unrelated, but now I see how they overlap, and I'm glad for it!

I am now the Class of 1952 Associate Professor of English and Education at the College of William & Mary in Williamsburg, VA, where I am jointly appointed to the School of Education and the department of English. I was also affiliated with the programs of linguistics and Africana Studies, and before that I served for 7 years as the inaugural William & Mary Professor of Community Studies. I direct the William & Mary Scholars Program and codirect the William & Mary Scholars Undergraduate Research Experience (WMSURE). In fall of 2017, I will become the North Hall Endowed Chair in the Linguistics of African America and Director of Undergraduate Research at the University of California Santa Barbara.

With support from the duPont Fund and the State Council of Higher Education in Virginia (SCHEV) College and Career Initiative, I work with admissions and development, and do a considerable amount of community outreach to students whom I hope to see at William & Mary and at UCSB and at other universities in 10 to 15 years. I want to be successful in the present, but my sights are set on the next generation and the generations after that.

My driving research statement is as follows: The quest to educate non-Standard English–speaking students from marginalized backgrounds has been a primary driving force behind both the multicultural education movement and the development of the field of sociolinguistics. These two perspectives, however, have not joined together as well as they could to address issues of language variation in multicultural education. Thus, I write at the critical juncture of sociolinguistics and multicultural education. My working premise supposes that only with an understanding of the principles and patterns of language variation in speech and writing can the multicultural education movement fully address why children from non-Standard English–speaking backgrounds often have difficulty achieving in schools.

Children innately know their language is not bad or wrong. Their families and neighbors speak the way that they do. But when they get to school, they are often told that they are incorrect and "speak bad English" and that therefore they are not as smart as other students. While recognizing the fact that literary language is crucial to academic success, we are working on ways to help children learn the standard language without teaching them their own cultures are bad. At the same time, we must work to include knowledge of all languages and cultures into what is taught, recognizing that all linguistic and literary traditions are just as valuable, not just the ones that are maintained in U.S. schools today.

My desire as director of the William and Mary Scholars Program is to build a program for students that will serve as a national model for nurturing the academic potential of high-achieving students from diverse backgrounds. As my career has progressed, I have seen the direct need to engage secondary and postsecondary English educators together in conversation. To do so, I have been a lead researcher on the Virginia Capstone English Project, the Senior English Seminar Academy, and the College & Career Readiness Initiative, statewide projects funded by the State Council of Higher Education for Virginia to align secondary and postsecondary English education goals. I also work with secondary English education students and doctoral students at the College of William & Mary on many of the same topics and themes that are covered in this book. The story we are now writing is our own.

Cheryl L. Dickter

I grew up in Middle Island, NY, in a middle-class community with a lot of racial and socioeconomic diversity. During middle school and junior high, I began to notice differences in the way that people from different groups were treated by teachers and staff at my school and even by parents: a teacher choosing me for a special award over my African American friend who had outperformed me; a parent expressing concern about his daughter going to a friend's house because it was in the "ghetto." Once I began noticing these individual events, I began to see instances of prejudice everywhere I looked. Through my senior year of high school, violence erupted between different groups, with racist comments and interracial dating as the catalysts of the violence. Our school hired security guards and installed metal detectors as I struggled to understand where this hatred came from, receiving little information from teachers and parents. Throughout this time, I asked questions about implicit and explicit prejudice that no one could—or wanted to—answer.

I chose to pursue my undergraduate degree at Randolph-Macon College, a liberal arts school in Ashland, VA. I picked this school for two primary reasons—it had a strong honors program that came with a substantial scholarship, and I would be able to start on the college soccer team as a freshman. Being a college athlete was a big challenge for me, particularly in my first semester when I was still adjusting to the transition from high school. Soccer was a commitment of at least 2 hours a day and involved travel to away games about once a week. I missed class and had less time to complete assignments, and I had to make up hours at my work-study job on campus. My grades were not where I wanted them my first year, but I did better my second year. One thing that I really liked about Randolph-Macon was the close relationships I formed with my professors. I took opportunities to see my professors outside of class in office hours and at academic events on campus. I even created a small venture in which I would go to professors' dinner parties to serve the meal and clean up so the host could focus on dinner! This was a great way to make some extra money, and more importantly to form relationships with faculty from different departments (and learn some gossip!). Although I intended to become a teacher with my degree in elementary education and psychology, my plans changed once I realized I could study the very questions about prejudice I had asked since I was in middle school. I took a course called the Psychology of Stereotyping and Prejudice and realized that there was a whole field investigating these questions and finding out answers that were so important in this world. Although there were not many research experiences available at my school because professors were focused on teaching, one of the professors I had gotten to know sought me out and offered me a position doing data entry for a study on racial prejudice and discrimination. Her research looked at how and why people choose—or do not choose—to confront racist comments. This was something that I was fascinated with in my own life, and here was a chance to actually study it! After working on this project with my advisor, she encouraged me to pursue graduate school, and I never looked back.

Like many students, I also dealt with a personal issue while discovering a love for research and while figuring out what I wanted to do with my career. My father developed cancer during my junior year and passed away shortly after. Particularly because my family was 8 hours away in New York, trying to complete an honors project, working part-time, and playing soccer while dealing with his illness and death was difficult for me. Sometimes having a loved one in the hospital or having to grieve a loss can make things like achieving a strong GPA seem unimportant. Having a support system got me through it—good friends offering shoulders to cry on and my research advisor encouraging me to think beyond graduation—and led me to an honors degree and admission to a great PhD program in social psychology at the University of North Carolina at Chapel Hill.

During graduate school, I became more independent in my research and began to investigate the implicit aspects of prejudice that had puzzled me in middle school. I learned neuroscience techniques and worked on examining how race is processed in the brain, and how this is related to prejudice, a topic that I continue to investigate today as an associate professor at the

College of William & Mary in the Psychology department and the Neuroscience program. Helping students find the research topic that has inspired questions in their lives and make discoveries in the lab to answer these questions is something that drives me every day. My research experiences were paramount to my success not only as an undergraduate student but also as a person struggling to understand the prejudice that her friends were facing. As a faculty member, working in my lab with students who have found their passion working on topics important to their own cultures and facilitating the transition to graduate school and an academic career is extremely personally fulfilling.

Hannah A. Franz

I am from Richmond, VA. Like many schools across the country, the public and magnet schools I attended operated under racially and socioeconomically segregated tracking, which impacted my academic and social life. A White middle-class student, I was pulled into gifted and advanced classes with most of my other White peers, but very few of my African American peers. The impacts of tracking by socioeconomic status were also notable to me, especially as I entered middle school, where a student's neighborhood seemed to determine placement in accelerated or honors classes. At the same time, I learned from my teachers how to cultivate empathy, and I also learned that great teachers come from all different backgrounds and speak all different language varieties.

I went to William & Mary for college and majored in linguistics. I took a course on Language Attitudes taught by Prof. Charity Hudley and learned about the implications of language variation in the educational injustices I had observed in school. We read widely on this topic in her course, and I continued to do so for my senior honors thesis research. With Prof. Charity Hudley as my thesis advisor, I undertook action-based research to design a website for teachers to learn about language variation, particularly African American English, and to find resources on how to use this knowledge to increase educational opportunities for African American English speakers. Because of my experiences with impactful teachers, I knew that teachers, if provided with the necessary tools, could do positive work for students from underrepresented backgrounds even within an unjust system.

My undergraduate research led me to further pursue sociolinguistics and education in a masters program at North Carolina State University (my undergraduate research was essential to my acceptance to and funding for this program!), to expand my knowledge of K–12 literacy at a masters program at the University of Pennsylvania, and finally to teach middle school reading back in Virginia public schools. I taught at a predominantly African

American school, designated Title I based on the percentage of students receiving free or reduced lunch. I learned from my fellow teachers, especially Letitia Frank and Larry Carr, and my students and their families how to communicate and collaborate cross culturally and how to focus education on community engagement. I also learned other, more difficult lessons. By the time I began teaching, U.S. public schools were notably different from when I had attended them. I found that what I had to teach was determined almost exclusively by standardized test preparation, and that different types of preparation assessments took up much of our instructional time. I began to think of other positions from which I could address educational inequities.

I returned to William & Mary as a PhD student in Curriculum Leadership at the School of Education. The most important aspect of my PhD work has been my involvement with WMSURE with Prof. Charity Hudley and Prof. Dickter. I work with college students from K–12 schools like the ones I attended and taught at. My research looks at how more students from schools burdened with educational inequities can be prepared for, access, and reach their full potential in college. I continue to have a focus on literacy and language, looking at how students can learn college literacies, even as K–12 systems demand that they instead learn standardized testing literacies. In writing *The Indispensable Guide to Undergraduate Research*, I aim to provide you, as a student, directly with access to these college literacies and the possibilities that they open up for you. In telling my story, I aim to provide you with an example of the opportunities that undergraduate research can afford for graduate schools and a career, and how research in college can lead you on a path to influence the lives of others. After completing my doctorate, I plan to continue my current work both by directly teaching students about reaching their full potential in college and by preparing future educators to teach for equity. Research will remain crucial to my work, as I explore the best ways to achieve these goals.

Who You Are!

Figuring out who you are is also research and a process of discovery that colleges are designed to help you with; make sure you take advantage of it. Greater academic equity is what we are striving for through this book . Our different academic experiences are varied yet overlapping. Not everything we have learned or that appears in this book is from knowledge that we have learned in a class or in school. Our academic experiences, our research experiences, and our social experiences are different, yet we converge at the point of writing this book and putting our ideas together. Many academics might have tried writing much of the information in this book from just a professor's perspective, but in this book we include many students who are

Activity 1.1: Discovering Others' Journeys

This is an exercise to help you get to know your professors and use them as resources. Ask a professor:

- How did you become a professor?
- What were some of your greatest successes along your journey?
- What were some of your greatest challenges?
- Was there a time when you failed at something you really cared about?
- What advice do you have about the research process?
- What freedoms might the next generation have that we haven't yet enjoyed?

going through the research process, and Ms. Franz, who has taught students in secondary school long before they come to college, to help you better understand your academic experiences as a process. Thinking not just about what you want to study but how you want to study it over time makes for a strong research process. Activity 1.1 is designed to help you start conversations with your professors about their academic journeys so that you can more fully discover your own.

In this book, we provide our working definitions of the terms we use to describe the social, cultural, racial, and economic categories that you may belong to, particularly as they relate to the college research experience. We also show where the terms are used in higher education. You are more than these terms. But our lens and research has shown us that culture and groups matter; your identity is an important dimension of who you are as a whole person and as a researcher. We also want to emphasize how the definitions and usage of the terms both vary among users and change over time. We encourage you to understand the historical and legal nature of the terms, but also seek to define the terms yourself, as self-definition is the true goal of a scholar. We make a distinction between the ways that terms are used to describe people with common experiences and the ways in which terms are used to describe the result of historical and/or contemporary inequalities that students face. Let's start with our most frequently used terms, explained in Textbox 1.2.

In Chapter 6, we identify and describe more specific pipeline challenges for students from groups who are underrepresented in higher education, including students of color and first-generation students, so you might already start to think of yourself as future professors and professionals with unique and valuable insights to contribute.

Textbox 1.2: Working Definitions of Social, Cultural, Racial, and Economic Categories

Underrepresented: We use the term *Underrepresented* to mean students that are underrepresented in the college population in general. We also use the term to describe inequalities of a population at a given college. For example, African Americans are underrepresented in higher education in general but are not underrepresented at historically black colleges and universities (HBCUs).

First Generation: A student who will be the first person in their immediate family to graduate from a 4-year college. More specific descriptors are students whose parents have attended a 4-year college but did not graduate and students whose parents attended or graduated from a 2-year college.

Low Income: We will explain this dynamic using concepts that demonstrate the economic demands that the families and students in our audience face that complicate the federal distinctions, such as the lack of capital in African American and minority families, even in the middle class.

African American: Used interchangeably with Black to describe students of the Africana Diaspora as a cultural and a racial descriptor. Students self-identify into the category. Every term for African Americans, Blacks, is both good and bad, and that paradox illustrates the struggle since before the time of the Racial Integrity Act of 1924, which sought to classify everyone as one race or another (no in between).

Latino/a/x: Used to generally describe students of the Spanish-Speaking Diaspora as a cultural and an ethnic descriptor. Specific terms used include: Latin American, Caribbean, Chican/a/x, Latin American, Hispanic, and Latino/a/x.

Person of Color: Used to describe people who identify as fully or partially non-White. The term is often used as a collective to express solidarity across groups but has the danger of also being used to lump groups together.

White: Refers to Caucasian individuals who are typically thought of as the majority racial group in the United States. Whites are well-represented in all academic areas because of current and historical privileges that come with being the majority group.

Other Social, Cultural, and Experiential Groups: While the text focuses on the interests of race, income, and educational experience, students in all groups will benefit greatly from the text. Such readers include Native American students, Asian American students, students with physical, emotional, or intellectual differences, and students who have experienced challenges not limited to but including: loss of one or both parents before or during college, homelessness or home insecurity, foster care, and refugee status. We will not have a focus on international students—students who did not attend high school in the United States of America. We encourage you to write similar books with focuses on populations that you are part of so that they ring true for others.

Gender Identification: A socially constructed term used to signify how an individual self-identifies with respect to biological and physiological traits. Scholarship on issues of gender and sexual identity has shown how dominant society determines gendered behaviors and ideologies. These ideologies include marginalization of alternative, gender-nonconforming practices as well as gender identification that is not cisgender, that is, that does not align with the sex one was assigned at birth. To be inclusive of all gender identities, we use "they, them, and their" as our default pronouns throughout this book.

Sexual Orientation: Used to describe who an individual is sexually attracted to. Sexual orientation is often conflated with sexual behaviors and sexual identity.

WHAT YOU CAN LEARN FROM THE REST OF THIS BOOK

This chapter has begun to provide you with questions and ideas about what you can do, know, be, and change as a scholar and a researcher. Chapter 2 (Get Started with Undergraduate Research: What, Why, and How), which focuses on what research is and how you can get involved, will further help you make connections between research and the transition to and from college with a focus on *what* research looks like, more reasons *why* you might want to get involved with research in college, and *how* to get started. Chapter 2 will describe the major ways in which research is categorized and why innovative research doesn't always fit neatly into the traditional categories. After learning about the whats, whys, and hows of research, Chapter 3 (How to Fit Research in with Everything Else: Time and Energy Management) will help you examine the *when*, so that you can start asking and answering questions to help you make the most out of your time and energy as you engage in research in particular and college in general. Chapter 4 (Research with Professors and Mentors) will take a look at *whom* you will be working with as an undergraduate researcher: your professors and research mentors. Chapter 5 (Writing and Presenting Research) will help you plan for sharing your research with others through writing and presenting. You will likely find that learning to write as a researcher will help you excel in college writing more generally. In Chapter 6 (Underrepresented Scholars in the Academy: Making a Way), you will learn about challenges for scholars and researchers from backgrounds that are underrepresented in higher education. You can start to ask and answer questions that will help you navigate these challenges and/or make a way for other researchers. We conclude in Chapter 7 (In Conclusion: Research in Action), continuing ideas from Chapter 6 to help you make a way for future generations of researchers and plan the next steps in your career as a scholar.

CHAPTER 2

Get Started with Undergraduate Research

What, Why, and How

As stated in the Preface, for the purposes of this book we are defining research as an inquiry or investigation that makes an original, intellectual, or creative contribution to a discipline, area, question, challenge, or theme. We encourage you to get started on pursuing research experiences early—starting to gain research experience during your freshman or sophomore year of college will pay off. The College Board even offers AP Research as a way to help you get started with the college research process in high school (College Board, 2016). The academic benefits (e.g., doing better in your classes) of research are greater when you start to participate in research early in your college career, as researchers Bauer and Bennett (2003) and Jones, Barlow, and Villarejo (2010) have found.

This chapter describes what a scholar or researcher is, the benefits of undergraduate research, and how to get started at your particular college. This chapter is not a how-to of the research process because there are many other books dedicated to that topic. Instead, we want to explain the hidden curriculum—the steps that have not yet been written down—of why and how to be an undergraduate researcher in the first place.

In Textbox 2.1 we pose a list of questions to help you get started on identifying your research interests, which can in turn help you as you look for professors and research labs to work with. When answering these questions, do not limit yourself to answers regarding academics or research—think about the answers broadly, in terms of your life both inside and outside of the classroom/research space.

So, why do we ask you these questions? Most people who do research started their path to conducting research with a question. The question often centers on something they were passionate about and that they wanted to find out the answer to. This topic could be something you question from your own life, your experiences with other people, something you learned in class, something you heard in the media, and so forth. Questions like these have helped our students become better learners. Scholar Nathanael Paige explains:

> Research has taught me to ask the right questions, both in academia and in everyday life. My analytical skills have been sharpened by research's demand for looking at an issue from multiple perspectives, while simultaneously being able to anticipate potential unexpected causes and outcomes. Thanks to research, I am also a better learner, because when reading textbooks I now have an improved understanding of what key connections to look for. Research improves the way you critically think, which consequently improves a variety of other things, like the way you communicate ideas and draw conclusions. When you participate in the production of knowledge, you learn to absorb new knowledge more strategically and purposefully, with the realization that everything is connected. I was prompted during a WMSURE session to think about what my passions were. I distinctly remember questions like, "What problem in the world do you want to fix?" Questions like this motivated me to relate my research to identity and perception influences.

We ourselves developed our research questions out of something in our own lives that made us question the status quo or fascinated us. When you study something that you are truly interested in, research can be exciting and fun! We all greatly enjoy what we do, which is important because in some cases conducting quality research is difficult, time-consuming, and even controversial. For example, as we described in our narratives in Chapter 1, we study issues related to inequalities, prejudice, and discrimination as related to social groups such as race. We chose to study this topic because each of us has experienced prejudice or unfair judgments from others and witnessed discrimination against others. It made us ask questions such as "why does this occur?" and "what can I do to change this?" Finding the answers to these questions led us to pursue our research in these topics and has kept us and our students excited about finding the answers to these questions. By thinking about the answers to these questions and talking about these answers with people around you, you can develop your own research inter-

Textbox 2.1: Identifying Your Research Interests

- What are you most interested in in the world?
- What makes you angry; what would you like to change?
- What would you like to create?
- What would you like to spend your time doing?
- What are major trends in academia, in industry, in the media, etc.?
- What are you studying in your classes that most interests you?

ests. Often, our students will tell us that they don't think that they can study something that they are passionate about because it isn't an academic topic appropriate for research. It is our experience, however, that most people can find some area in which they conduct their research based on their passions. For example, did you know you can do research on hip-hop culture or video games or disaster relief efforts? Students we work with are researching each of these topics. We promise you that you will be able to find an area of interest and do research on any topic. Scholar Brian Anyakoha explains the benefits of finding a research topic according to his goals and interests:

> The way I see it, anything you do in college and any choice you make should reflect you: who you are and what your goals/interests are. It should also reflect experiences, in class and life. I have two passions as a student at the College: science and music. Those are two vastly different fields, but through research I was able to fuse them into a project idea I am in the process of getting off the ground, and which fits nicely under the domain of my Neuroscience major. This all came about after I took my freshman seminar, "Chemistry of Emotion," and did a presentation on neural circuitry that correlates with the emotion we feel when we hear our favorite songs. It was actually this class and the work I did for it that propelled me into my major, and away from my intended Biology major. And I can't say I'm disappointed as of yet! Why? Because I chose to work for and toward something that was near and dear to me and that I would be genuinely interested in. The days of doing things half-heartedly, or because you have to, begin to go away in college when you realize that you can do and learn things for you and not because someone told you to. In this way, research becomes a very liberating and rewarding experience.

Once you narrow down your area of interest or research questions, either in a class or on your own, you can start trying to pursue research opportunities. A good way to get started is to browse the websites of the faculty members at your university. Chapter 4 can give you strategies to use and questions to ask professors who might provide research opportunities.

You can get started even if you are not yet in college! Check out the options that you are interested in now to plan for the future. Start checking professors' websites to see what interesting research they are doing. During prospective student events or your classes, ask a professor what kind of research they do to get an idea of the work that is going on at your school.

STUDENT VERSUS SCHOLAR

Many books and articles that focus on college success aim to prepare students to obtain good grades, overcome obstacles, and graduate in good standing prepared for a career. The focus of this book, however, is to help you *excel* in the university setting. For many of you, it is likely that you will work hard and succeed so that you achieve strong grades and learn much from your courses and activities. We want to stress in this book, however, that although it is important to be a successful college student, we think that it is extremely important to have the goal to excel in college and to reach your highest academic potential. By highest academic potential, we mean learning outside of and across your classes, asking questions that haven't been answered yet, and creating new knowledge. Scholar Dondré Marable describes this difference:

> You begin learning things on your own in a way that does not compare to sitting in a lecture hall for 50 minutes. You gain practical hands-on knowledge that, in many ways, is applicable to the real world. Instead of learning about a topic and regurgitating the facts for an exam, you take the concepts you have learned and apply them in such a way so that you may potentially contribute to the world's knowledge domain.

As we discussed in Chapter 1, we think of this difference as the difference between being a student, in which you learn important information and graduate with a degree that should help prepare you for the world, and being a *scholar*, in which you reach your greatest potential and are well prepared for the graduate schools, professional schools, and workplaces that are the best for you and your goals. Part of what we think will make the difference between being a student and a scholar is participating in research experiences. Scholar Danielle Weber explains how research helped her transition from student to scholar:

> When I first became involved in research, I was excited by the topic, but daunted by the process. I saw research as a long, tedious process, and I imagined my role to be no more than as an acknowledgment tucked away on some future report. After some time with this mindset about research, my advisor alerted me that the data collection phase of the study was ending and I could start to run analyses. I was caught off guard—I was around for the start of this study, but I never guessed I would see it end. When my advisor taught me how to process and analyze my data, I was excited; this wasn't homework or something to be

tested on, but a skill to help me present my work. Even when I started writing the manuscript, I initially treated it as a class assignment, just working for the approval of my advisor. After rounds and rounds of edits of the manuscript, I realized that her suggestions weren't about my syntax or phrasing, but about how to convey the importance of my research to an audience. This wasn't simply a report of what I did or what analyses I ran, but the very reason why I decided to do it in the first place. Even the precise and time-consuming writing of the statistics and references didn't feel tedious because it wasn't for a grade, but to make a final product in a peer-reviewed journal. The day my manuscript was accepted, I remember feeling overwhelmed by the prospect that my study may inspire future research. I was proud of the accomplishment, but I did not see my research as *finished*—I thought about numerous follow-up studies and where the field might go. When I first started research, I had never considered a research career. Now, several years later, I am about to start a PhD program to pursue a career in research. When that manuscript was accepted and the journal's editor accidentally referred to me, an undergraduate student, as "doctor," I remember thinking "I like the sound of that!"

RESEARCH HELPS YOU BECOME A SCHOLAR

So what's the big deal about research? We obviously think that research is great, since we chose to write this book about the topic. But how does working on research in college help you obtain the status of a scholar compared to that of a student? Although you will learn about issues and methods in academia and in the world in your courses, research allows you to apply concepts learned in your content courses to finding out answers to the questions that you have about these issues. For example, you might learn about concepts related to prejudice in your psychology courses, but working on a research project may allow you to explore more specific questions that are important to you regarding this topic. Although another researcher may have conducted some previous work on your topic of interest, you bring a unique perspective to the topic based on your own experiences and opinions. And you also learn about the specifics of the research process.

Let's look at the academic research on this topic. For this research, colleges and universities conduct large surveys in which they ask college students to report whether they have conducted undergraduate research and then they measure a variety of outcomes in students, such as GPA, involvement in school activities, enjoyment of classes and major, as well as demographic information such as gender, race, and type of high school attended.

Looking at many studies conducted at various institutions, we can get an idea of how students from a variety of backgrounds and schools who participate in research differ from students who do not participate in research. This research has shown that undergraduates who conduct research do better in school and are more likely to graduate on time than those who do not conduct research. For example, a study at the University of California-Davis conducted by Jones, Barlow, and Villarejo (2010) reviewed the academic progress of 6,834 students majoring in biology. They collected information from the admissions office on GPA and SAT scores, and they reviewed the grades of the students during college and examined how long they took to graduate. The researchers found that, holding the above variables constant, compared to students who did not conduct research, those who conducted research in any subject during their junior or senior years were almost 15 times more likely to graduate and those who conducted research for more than 2 years were 900% more likely to graduate! Other studies like this have found similar effects of undergraduate research at other universities, both small and large, and in other subject areas.

Why might these positive impacts occur? Faculty who were interviewed about undergraduate research said that students involved in research are more likely to come to their office hours and therefore gain a network of people to help them succeed in school. Students who conduct research also tend to be more involved in their campus communities (Kuh, Kinzie, Schuh, and Whitt et al., 2010). Their research also revealed that these benefits are even greater for students who enter college with less academic preparation (i.e., attended high schools with fewer honors/AP/IB options or had a lower quality of high school teaching), underrepresented minority students, and students who are the first ones in their family to attend and/or graduate from college (first generation). For example, Lopatto administered a survey in 2007 to 1,135 undergraduate students from 41 universities. Although the research revealed that most students who completed a research experience reported gaining knowledge about research, African American, Hispanic, and Native American students reported higher learning gains, such as being better able to work independently, overcome obstacles, understand the research process, and become part of a learning community. In addition, 98% of the students surveyed reported that they liked their research opportunity enough to recommend it to a friend. Importantly, the benefits of undergraduate research do not vary as a function of the institution you attend; that is, whether you attend a large, small, or medium-sized school, participating in research can yield a multitude of benefits, including better grades, increased professional skills important for college and career, and personal development. Because of all this research showing the many benefits that students can get from undergraduate research, large amounts of money are being invested by federal and private agencies to support undergraduate research (e.g., National Science Foundation [NSF]). There is a national organization committed to promoting and supporting undergraduate research called the Council on Undergraduate Research (CUR). CUR recognizes the benefits of undergraduate research, many of which we discuss in this chapter, for both students and professors. CUR provides resources and money to educate faculty about how best to support undergraduates pursuing research under their leadership and provides funding and information for undergraduates interested in conducting research with a professor. Indeed, the number of undergraduates participating in research, particularly in the sciences, has been increasing over the past decade (Laursen, Hunter, Seymour, Thiry, & Melton, 2010).

There are many additional reasons why becoming involved in the research process will have multiple benefits for you. First, doing research as an undergraduate student gives you insight into what research is all about. It is likely that you have heard the term "research" discussed in your classes, and many of you may have already conducted a research project as part of your classes. Doing research outside of the classroom can offer you an inside perspective as to what research really is. Whether you are working in a lab, reading archival material, or giving a performance, you will learn the particulars of the research process in a particular area of study. In our experience, it is only when students get involved in research courses or opportunities that they truly understand what research is, as it is a lived experience. Such experiences and courses may help you identify whether research is something that you would like to continue to take part in. We find that, after beginning to work on a research project that is personally interesting to them, some students find their passion for research and realize that they want to continue pursuing research at the undergraduate and graduate level and/or as a career unto itself or as part of their future job responsibilities. Indeed, faculty working with undergraduate student researchers described in an interview for a research study that their advisees made gains not only in research but also experienced personal and professional gains such as a clarification of and better preparation for career and/or graduate school goals (Hunter, Laursen, & Seymour, 2007). In interviews with students, those who had conducted undergraduate research reported that preparing and presenting their research to others increased their confidence in their professional ability. Undergraduates who attended professional research conferences (see Chapter 5) felt a great sense of accomplishment and pride in their work and reported establishing professional relationships that would enhance their career success. Regardless of whether or not you develop a love for research, research can have many other personal and professional benefits.

YOUR COLLEGE ALREADY SUPPORTS RESEARCH AND RESEARCHERS

One of the reasons why research at the collegiate level is emphasized by us and by other academics is because your college or university already supports the pursuit of research inquiry. In fact, many professors at colleges and universities conduct research in one form or another. What type of college you attend may determine whether or not research is pursued by professors there and how much time is spent on it; Chapter 4 describes professors' main responsibilities by type of college. For example, at a small school that is primarily focused on undergraduate teaching and at most community colleges, most faculty spend more time on teaching than they do on research. At these colleges, it may not be feasible for professors to conduct much research because of the large amount of time they spend in the classroom or because of a lack of resources to support research. Still, like research-focused colleges, many teaching-focused colleges have programs for undergraduate research. At medium-sized and larger universities and at universities with larger endowments, conducting research is part of a professor's job, and these schools tend to provide faculty with more time and resources for research. The size of the college does not necessarily determine whether research is a focus for your professors; some professors at large schools, for example, are responsible only for teaching. So how do you find out how much research is conducted at your school? Contact professors—ask them each if they do research and if so, what type? Even if they don't currently do research, they can tell you about the research culture at your school. Go to your professors' websites: Many of them will have information about their research there. You'll have an excuse to talk to your professors about something they are interested in (they'll be impressed you took the initiative to look at their sites), and you will also get an idea of the type of research that goes on at your school. Another option is to see if your school has an undergraduate research program or center. Some schools have one undergraduate research program, and other schools have different programs for different academic departments or majors. The undergraduate research program director may be a good resource, especially if you're having trouble finding out about research on your campus.

Professors at colleges around the world, for the most part, love to work with students on research. For us, the authors, working with students is one of the reasons we love our jobs so much. As you start to learn why you want to work on research with a professor, it's also important to learn why professors might want to work on research with students. As a unique individual, you can give us insight into our research from a different perspective than we might have. That's why diversity of backgrounds and perspectives is so valued. You provide us with energy and excitement as you're learning about the research process and you help us get our research done. We can see you grow as new researchers as you gain experience and start having your own research ideas. Undergraduate research programs improve the quality of the college: The best students are more likely to attend and stay at schools with strong research programs, and strong research programs also attract and retain good faculty. Professors who work with undergraduate students also qualify for special grants that can provide them with money to aid their research. Major organizations recognized for their commitment to education at the collegiate level, such as the Association of American Colleges and Universities (AAC&U), have identified undergraduate research as an important learning practice that universities should embrace (Kuh & O'Donnell, 2013). To encourage undergraduate research, many universities have developed undergraduate research centers or programs, many of which provide scholarships or grants to help you conduct your research project. Check out the website for undergraduate research at your school, go to the research center to get more information or, if you are having trouble finding that information, talk to a professor about the resources that are available on campus.

RESEARCH ALLOWS YOU TO DEVELOP SKILLS AND CONNECTIONS THAT WILL HELP YOU ON YOUR CAREER PATH

Because conducting successful research involves a number of different skills, working on a research project will help you sharpen your existing skills and develop new ones. For example, successful research necessitates the ability to think critically about a research area to the extent that you are able to understand a body of literature that has already been written on a topic and identify how this previous work fits together. Research often involves creativity in different modes: You may see a research topic in a different way than others see it, or you may share your creative research in the performing arts or in your writing. Research also requires problem-solving abilities in a number of different ways.

Finally, because you will be working with previous literature and presenting the results of your research project, you will refine your reading, writing, and speaking skills. These are all tools that will make you more successful at the college level, the graduate level, and in your profession. Universities and employers know that research experiences give students a high level of academic challenge, promote active collaborative learning, and provide an enriching educational experience. They understand how these experiences can translate into academic success or professional productivity. Thus, research experience on your resume or CV (curriculum vitae, your academic resume; learn more in Chapter 5) will increase your chances of getting into a graduate program and landing a great job. For example, in a study that followed undergraduates studying science at a large university, compared to students who did not conduct research, those who conducted research got a job faster and were more likely to get a job related to their academic interests (Kinkel & Henke, 2006).

In addition to the academic and professional benefits that research offers undergraduate students, research experience also can help you develop personal skills that may not be taught in the standard college curriculum. Numerous surveys of faculty and students who have taken part in undergraduate research experiences at different universities throughout the United States have demonstrated that, over time, students who do research develop greater self-confidence, show better independence in work and thought, develop a strong work ethic, learn ethical standards, develop improved communication skills, and feel a greater sense of accomplishment. For example, one study asked 320 biology and chemistry majors from 21 states about their research experiences. Students reported that their research experience provided them not just with laboratory experience but also improved their problem-solving skills and self-confidence (Mabrouk & Peters, 2000).

As scholar Katie Mitchell explains, the skills that she learned from conducting research landed her her first-choice job:

> The act itself of conducting research is just as marketable as the content of that research. None of the jobs I applied for had anything to do with the subject of my research, but my research experience made me a very competitive candidate. When I was invited to the final round of interviews for a position at my top choice company, a well-respected consulting firm, I asked my recruiter what the company had liked about my resume. It wasn't my GPA, the number of clubs that I participated in, or the awards that I had won—she told me that they were impressed by my experience managing research projects. My interview with a prospective manager focused on my experience collaborating on research projects from conceptualization through completion. This understanding of process is one of the greatest gifts of undergraduate research. You learn how to ask questions and then find a way to answer them based not on conjecture or popular opinion, but on a statistical approximation of reality. You learn that that's hard to do. You learn that data are messy, and that in between the generation of an idea and the publication of results there is a lot of tedious, frustrating, important work. You make a lot of mistakes, and you learn how to make fewer mistakes the next time. These lessons from research are less tangible than the results of a study, but just as valuable, and employers recognize that. I got the job.

Undergraduate research also improves students' levels of professional advancement in that it provides opportunities for publications and presentations, helps build a resume, and can help students develop relationships with professionals who can assist in academic and nonacademic employment. Research studies demonstrating these impacts have typically been conducted on large samples of students from all over the country. Russell, Hancock, and McCullough (2007) sent online surveys to about 15,000 people, including 4,500 undergraduate students and 3,600 faculty and graduate student mentors between the years 2003 and 2005 to better understand the professional benefits that research could have. Students who conducted research for more than a year reported stronger expectations regarding their obtaining a PhD (30%), a research-based degree, than those with no research experience (8%). There are opportunities to present your research at a conference where you will network with other individuals whom you can draw on for professional advice and who may be able to provide opportunities for graduate school or jobs. You may also work on publications in which your work gets published in an academic journal or book. Having these presentations and publications on your resume will impress anyone who reads it, because employers realize it takes professionalism, hard work, and excellent time management to present research at a high level while you are in college (see Chapter 5 for more about presenting your research). Undergraduate research can also help you understand the demands of a career and increase your knowledge of the professional world. Indeed, undergraduates who participate in research have a stronger sense of their career and graduate school plans beyond graduation (Bauer & Bennett, 2003).

Participation in research experiences may allow you to collaborate with other students who have research interests similar to yours or who are also going through the research process. Together you will help each other on a research project and you will have the opportunity to learn from more senior students and eventually to mentor newer students. In a sample of more than 2,000 college sophomores at 23 colleges across the country, students who conducted collaborative undergraduate research with their peers were asked to complete scales developed by researchers to assess a variety of personal and academic variables. In this study, undergraduate research experiences that involved collaborative learning with peers were related to students' own reports of how much they understood and appreciated science and the arts (Cabrera, Crissman, Bernal, Nora, & Pascarella, 2002). In addition to working with peers, you will also work closely with a professor. Depending on the size of your college or university, you may not have a lot of close contact with your professors through your classes. Research experiences with a professor will likely necessitate spending a lot of time working together in individual meetings or in small groups. The research advisor will likely act as a mentor for you, which is important as students who have mentors as undergraduates make a better transition to college, show better problem-solving and decision-making skills, and are happier with their college experience than students without mentors. A study of more than 300 professors and undergraduate students found that more than 20% of students felt that research experiences made them better at establishing and succeeding in collegial, working relationships with their professors and peers (Hunter, Laursen, & Seymour, 2007). Having a mentor-mentee relationship will give you a chance to know your advisor on a deeper level and to establish a close professional relationship. We maintain close relationships with our research advisees even after they have graduated, as described in Chapter 4. We see many of them at conferences and still have some of them come back on campus and visit us years after they have graduated. Sharing in the success of our students in their careers but also in their personal lives is one of the most rewarding parts of being a professor. For you, getting to know your advisor this well can establish strong, lasting relationships that can help you professionally throughout your career and in your personal life. In addition, securing a close relationship with your research advisor will allow you to have a professional reference for life! Your research advisor will be able to write you a recommendation letter for graduate and professional schools or serve as a reference for jobs to which you will apply in the future. Because your advisor will see how intelligent, motivated, organized, and hard working you are (to name just a few qualities) based on your research project, they will be able to give potential schools or employers a well-rounded idea of why you would be the perfect person for a specific position. They'll be able to back up these assertions with examples of your success in your work. Having an endorsement like this can be very important, as most other professors who will write your letters will most likely only be able to describe your performance in your classes. For example, they may be able to say that you are intelligent and studious, but they may not be able to speak about other skills that graduate schools and employers may see as being vital for success.

RESEARCH CAN EARN YOU MONEY, ACADEMIC CREDIT, AND A TANGIBLE SCHOLARLY CONTRIBUTION

In addition to the learning experiences you will have and the relationships you will develop, there is another important reason why we encourage students to conduct undergraduate research: You can get money and academic credit for researching issues that you are interested in! That's right—you don't have to just volunteer your time working with your research advisor;

you can also get tangible rewards for doing so. You can earn college credit for working on a research project during the school year and you can get a salary for working on research, in most cases during the summer.

Most students also feel pride from conducting research that results in the creation of a project that can contribute to an area of study. For example, your project may result in a scholarly presentation, an academic paper, or a performance that you can share with others. Telling others about what you discovered, learned, or did during your time as an undergraduate researcher can be personally fulfilling, and you can also feel a sense of accomplishment that your project has contributed to your area of interest. Maybe the findings of your research will be taught in future undergraduate classes at your own school. We have had undergraduate students who have been published in academic journals, have given presentations or performances to the top experts in their fields, and have made important contributions to scholarly books. We remember the first time we saw our names in print listed in a conference book, at the top of a research article, or on a book, and these were all very moving experiences. Showing our family and friends a tangible product of all our hard work wasn't too bad either.

WHAT RESEARCH LOOKS LIKE WITHIN AND ACROSS RESEARCH AREAS AND DISCIPLINES

Research is generally categorized into the humanities, social sciences, or natural sciences. We'll talk more about each of these categories below, but first we want to emphasize that an important component of all disciplines is the review and dissemination of information learned from the pursuit of knowledge. Research can take place in a laboratory, a museum, a library, or a community. In our description of research so far, we've been very careful to broadly define research, how it is conducted, and what the products of research are. The reason for this broad definition is that research can vary widely among different disciplines. While much research takes place within one discipline, interdisciplinary research combines approaches from existing disciplines into new approaches to research questions. Your research may even lead to the creation of a new discipline or interdisciplinary method or theory. We will next turn to explaining a little more about what research looks like across the various disciplines. Before getting into each broad area of research, it is important to point out that not all disciplines fit into just one area of research. In many cases, the research methods that are used by a subdiscipline are what dictate the classification of the area. In other cases, history dictates notions of disciplines. At other times, scholars within a particular area will dispute the classification of a discipline. For example, many researchers who study linguistics, the scientific study of language, view themselves as participating in all three disciplines: the humanities, social sciences, and natural sciences.

Humanities

The humanities are disciplines that study human expression, thought, and culture. The humanities often involve the interpretation of previous research or art within a historical and cultural context. Research in the humanities tends to use historical, interpretive, and analytical methods to make connections, examine meanings, and explore contradictions. Humanities research is often conducted by an individual professor or professors researching previous works in order to write a scholarly work such as a book or an article. Academic disciplines that are considered humanities are ancient and modern languages, religion, literature, philosophy, art, and music.

Social Sciences

The social sciences are disciplines that examine relationships among individuals in a society and in their social and cultural groups. Social scientists can use the scientific method to understand society or can use more of a theoretical or interpretive perspective. Many social sciences use both quantitative techniques, which use statistical or computational measures, and qualitative techniques, which rely on data from spoken or written words. Some of the social sciences are economics, political science, sociology, anthropology, archaeology, history, law, psychology, and linguistics.

Natural Sciences

Disciplines in natural science aim to describe, predict, and understand objects or processes of the physical and biological nature using scientific methods. Most of this work entails using observational measures and empirical evidence. Examples of natural sciences are math, physics, astronomy, chemistry, Earth science, biology, and engineering.

Interdisciplinary Approaches

Many research areas have groups of researchers that are now focused on interdisciplinary approaches to research questions. The STEAM movement, which adds an emphasis on the arts and creativity to science, technology, engineering, and mathematics, is one you may be familiar with from your

K–12 experience. Public health, urban planning and architecture, business, and education are all good examples of popular research areas that value interdisciplinary approaches.

Research by Underrepresented Scholars

Many topics and questions that were not brought up in high school—or that may even have been taboo (Ayers & Ayers, 2011)—are welcome and of utmost interest in undergraduate research. They are understudied topics, and we need you as the next generation of researchers. As we mentioned earlier in the chapter, we study topics related to inequalities, prejudice, and discrimination, which are topics that some see as difficult to discuss or study. We study these areas because we believe these topics are important to our society and we are okay with pushing the boundaries of others' comfort levels in order to make important discoveries and to implement change. Our unique perspectives based on our backgrounds have led us to ask different and often difficult questions, and our research has helped us to answer these questions. You have a unique background, and it may lead you to want to discover things about yourself, your family, your neighborhood, or topics in the world that you are curious about. Conducting research on subjects about which you are passionate can lead to the best work. As an underrepresented student, you may have a different perspective from that of your professors and the existing researchers in the field. Because most research necessitates extending previous work into novel areas or starting a completely new line of research, being underrepresented can lead to successful research ideas. We have had students ask questions inspired by their life circumstances that have led us to pursue new research questions and to make a new contribution to the field.

There are initiatives to include underrepresented scholars in all forms of research. There are special programs and initiatives at individual universities, funded by foundations such as the Ford Foundation and the Mellon Mays Undergraduate Fellowship Program and by state and federal government organizations such as the National Science Foundation and the McNair Scholars Program, to promote research of different types by underrepresented scholars. (See Chapter 6 about underrepresented scholars in the academy for more information.) The perspective of underrepresented scholars is very important because it brings entirely new ideas to research. In addition, there is often an ethos of transformation and "paying it forward" among underrepresented scholars that urges us to contribute greatly to scholarly organizations and universities (Ellison & Eatman, 2008). We tend to do more than our share of mentoring and advising students on campuses, even at research universities. Having a strong background starting

TEXTBOX 2.2: BLACK ACADEMIC AND PROFESSIONAL ORGANIZATIONS

- National Association of Black Educators
- National Association of Black Engineers
- National Association of Black Psychologists
- Black Doctoral Network Association
- Association of Black Sociologists
- Association of Black Anthropologists
- Society of Black Archaeologists

from your undergraduate years will ensure that you are able to balance all of your desires and demands from the beginning of your career. We want to make sure that you have the technical and theoretical aspects of research down long before you take a position in academia or industry!

Underrepresented researchers in some areas have specific societies so that our academic and social interests are supported. For example, some Black academic and professional organizations start with "National" because "American" usually indicates primarily White organizations. Examples are listed in Textbox 2.2.

Empirical data from the National Center for Education Statistics (NCES) study (Aud, Fox, & KewalRamani, 2010) and National Science Foundation (NSF) data reveal what underrepresented scholars currently study and where there are challenges with representation. The Institute of Education Sciences (IES) data reveal that first-generation students are heavily represented in business. In another example, African American and Latinx students and women are most represented in STEM (science, technology, engineering, and mathematics) in the biological sciences (U.S. Department of Education, National Center for Education Statistics, Integrated Postsecondary Education Data System, 2013). Individual colleges and universities have data about the gender and racial backgrounds of students and researchers by major and discipline; for public universities you can request the information by law.

Faculty diversity in research is even more limited. Trower and Chait (2002) explain that policies and values at research institutions must change in order to provide for more inclusion. Table 2.1 from their research emphasizes ways in which the research academy is changing.

Underrepresented scholars, because we bring perspectives that are new and particularly unique to historical research contexts, have made great contributions to all types of research. Scholars from underrepresented backgrounds have been instrumental in breaking down traditional boundaries in

Table 2.1. How the Research Academy Is Changing

New View	Old View
Transparency of the review process assures equity.	Secrecy assures quality.
Merit is a socially constructed, subjective concept.	Merit is an empirically determined, objective concept.
Cooperation is better than competition.	Competition improves performance.
Research should be organized around problems.	Research should be organized around disciplines.
Excellent teaching and advising should pay off.	Research is the coin of the realm.
Personal life matters; balance is important.	Separate work and family.
Faculty have a collective responsibility.	Faculty thrive on autonomy

Source: Trower and Chait (2002)

scholarship based on disciplines (see Textbox 2.2). Intellectual approaches including African American/Africana and Latinx studies represent cross-disciplinary and interdisciplinary approaches to questions across space and time. Orlando Taylor explains in Ellison and Eatman (2008) that the reason that many underrepresented scholars get into academia in the first place is to help educate the next generation and to tackle societal challenges. In Activity 2.1, we give you some questions to help explore the research of underrepresented scholars on your campus.

OPTIONS FOR UNDERGRADUATE RESEARCH

We hope that by now we have convinced you that undergraduate research is worthwhile for you to pursue and have given you an overview of what research looks like in different disciplines. So how do you go about pursuing a research project at your university? There are usually many different options for conducting research with a faculty advisor. These options will differ based on your school, so make sure you check with your school's undergraduate research center or search for online resources. Asking a professor how to find out about your options is also a good way to figure out what types of opportunities you'll have.

Before we talk about the options that you can find on and off your campus to conduct research, we will answer some general questions about the research process that we often hear from our students:

ACTIVITY 2.1: EXPLORING RESEARCH BY UNDERREPRESENTED SCHOLARS

- On your campus, what do faculty from underrepresented backgrounds teach?
- What do they research?
- Do underrepresented faculty on your campus consider themselves part of one or many disciplines and research approaches? What methods do underrepresented faculty use? Who supports their work financially and intellectually?

1. *Do I have to come up with my own research idea?* Not usually! Most students will start out working with a professor on an existing idea or project or will help continue a professor's research on a new project or line of research. If you find a professor whose work you are interested in, you can start out by helping with projects that are conceptualized by the professor and/or other students.
2. *What if I have my own research idea?* Great! Many students, particularly those who have already gained some experience working with a professor on research, will generate a research idea that is of interest to them. In this case, you can propose your idea to a research advisor and see what they think about the quality of the project and the feasibility of conducting the research. Before you bring up your ideas for a research project, be sure to thoroughly research the topic and have a strong understanding of the relevant background literature (see Chapter 5 for tips on finding previous research on a topic). In most areas of study, your project will need to be unique, so you should make sure that no one in the field has already done an identical project or something very similar. If you do find that this is the case, don't get discouraged—you might be able to use this past research to inform a new project or go in a different direction. If you end up doing a research project that is based on your own idea, the parameters of the final project will often involve a negotiation between you and your advisor. This negotiation occurs partly because your advisor has expertise and interest in certain areas of research and may not be able to provide you with adequate support on your specific proposed project. In this case, you can talk about designing a project that would fall into your area of interest and that would also be something your advisor is interested in and capable of supporting. In addition, your advisor will likely have some suggestions on ways to improve your proposed project and will work with you to strengthen your initial idea into the best research possible.

3. *What if I don't have previous research experience?* The answer to this question will depend on the research advisor. Some professors require you to have taken a certain number of courses in the discipline or with them as the instructor before they will consider you for research—this is one of the reasons we encourage you to talk about research with your professors early, even in your first year of college. Other professors require no prior experience but decide based on your enthusiasm and interest in the topics they study. Some may require an application. There will also be professors who are somewhere in between these approaches. You should check with your potential research advisor and understand what the requirements are.

Whether you develop your own research project that is based on your own ideas or work on a faculty advisor's project, you can usually do research either as a volunteer or for course credit. As we mentioned earlier, most professors on your campus will be involved at some level with research and will benefit from having undergraduate students involved with their work. As an undergraduate researcher, you might be responsible for finding sources and literature on a research topic, summarizing this literature in written and/or oral form, collecting data, organizing and analyzing data, writing or presenting results (see Chapter 5), and so on. In most cases, you will be learning new skills and methods that you can utilize for the projects on which you will work that are specific to your discipline. Many institutions offer course credit within a major in exchange for working on research with a professor. Research also happens within courses. Many of you will complete research as part of one or more of your classes. You will most likely be responsible for reading and summarizing the literature on a particular topic and proposing or conducting a research project. Sometimes you will be able to choose a specific topic within the context of the course, while other times you are assigned a topic. There are also some classes in which you will do a project by yourself, and there are others that will require that you work with a group.

If research happens outside of an articulated course, you will come to an agreement with your advisor about how many credits you will take and what your advisor's expectations are for you to earn academic credit. You should be able to count this academic credit toward your major and your graduation requirements. You should consider this credit to be equivalent to a regular course that you take at your school. That is, if you are doing research for credit, you will need to fulfill the expectations laid out at the beginning of the semester in order to receive a good grade in the course at the end of the semester. If you do not want to have that level of commitment, volunteering to work with a professor can be a good way to get involved

Activity 2.2: Explore Undergraduate Research Opportunities

Explore the different opportunities for undergraduate research on your campus. What are the availability and pros and cons of the following types of research for you on your campus?

- In and across classes
- Volunteer
- Paid
- For credit
- Capstone project or honors thesis

with research without the pressure of academic credit. Working as a volunteer, however, will also require an agreement between your research advisor and you in terms of what you will be responsible for doing. Activity 2.2 is designed to help you explore the undergraduate research opportunities on your own campus.

Throughout this section we have been talking about working with professors on research projects. At some institutions, particularly those in which there are graduate programs, it may be that you are not reporting to or working directly with a professor. Instead, many professors at universities with graduate programs have their undergraduate researchers work closely with their graduate students. The graduate students in this case may serve as mentors more so than the professor. Often graduate students work as liaisons between the undergraduates and the research advisor. Although this model works well in some cases, this may not be a model that you feel works for you. When discussing the terms of your involvement in research, make sure that you understand whom you will be reporting to and who will be responsible for teaching you research skills. At schools without graduate programs, it may be the case that you will report to a senior undergraduate student rather than the professor. So, again, you should clarify everyone's role in the research process before you make your decision about whether you want to be involved.

Once you have obtained research experience and realized that research is for you, you may be ready to conduct more independent research, which can take the form of an honors project during your senior year of college. An honors project will give you an opportunity to spend a period of time (usually two semesters or a year) leading a project on which you are the primary researcher. As the primary researcher, you will be responsible for conducting most of the research, under the leadership of your faculty research advisor. Although the requirements at individual institutions will differ, in general you will plan and conduct a research project, which will culminate in a final paper or performance. You will then present the findings of your project to a group of people, usually a small group of faculty from inside and outside your department. During this presentation, you will demonstrate your knowledge of your project and the subject area while having a discussion about the strengths and weaknesses of your work and the implications of your findings.

Most of the research described in the previous few paragraphs has focused on research opportunities throughout the academic year. There are opportunities beyond the academic year that may also be of interest. In particular, the summer is a great time to work on a research project. Many schools will provide funding for students over the summer so they can get paid for working on a project. If you don't want to apply for a summer grant at your college (or if you don't receive one), look for schools that offer other incentives to stay on campus to conduct summer research, such as free housing. Whatever option you choose, you will have the opportunity to spend all your time on research without the distractions of classes. If you greatly enjoy your summer of research, there is a good chance that graduate school might be right for you. After all, most graduate programs focus on research, and that's what you'll spend most of your time doing! Working full-time on research over the summer can also further facilitate relationships with your professor and other students and help you develop stronger research skills. If you don't want to commit to spending an entire summer working only on research, you can work on research to varying degrees. Check with your faculty advisor to see if there are volunteer or academic credit options over the summer where you can work part-time on research. This option can free up other time to take summer courses or to get a job or internship.

You don't need to stay at your university to conduct research over the summer. There are great programs around the United States and also internationally that you can attend to get more research experience. Off-campus research is a particularly good option if there is not a faculty member at your college who does the research you are primarily interested in. In many cases you can compete for funding to attend other institutions and get paid for participating in their summer research programs. Many colleges, universities, companies, and other institutions (e.g., University of Michigan Undergraduate Research Program, The Ohio State University Summer Research Opportunities Program) offer summer research programs with varying levels of support. You should ask your faculty research advisor and search other colleges' websites to see what paid research opportunities there are for you. Spending a full summer focused on research can be a very rewarding experience and a great way to find out if research and/or graduate school are right for you. Off-campus experiences are a great way to broaden your research experiences; get exposed to another institution and geographic area; get to know other researchers and students in your field; and learn more about your discipline. We recommend that you apply to many different programs, as they can be competitive. The deadlines for these summer programs also tend to be early—usually between January and March—so you should be researching programs late in the fall semester and during winter break.

Internships are also a great way to get some professional experience. They can give you some insight into a potential career you might want, allow you to network and establish connections with individuals in your field, and give you the experience that potential graduate schools and employers look for. As with the research opportunities we've described above, a lot of these are paid opportunities or you can earn academic course credit for them. The career center at your institution can help you find internships that fit your academic goals, and you can also search the Internet for internships in your particular discipline. The department administrative assistants at your school also are a wealth of information about research and internship opportunities, as a lot of information goes through them. Many internships are offered during the summer. You can typically find internship opportunities around your college town, your home, or at an institution to which you would need to travel for the summer.

Now that you understand the benefits of research and how to seek out research opportunities, it is important to think about how to fit in undergraduate research with all of the time commitments you already have as a student and as a person. Chapter 3 is dedicated to prioritizing your daily, weekly, and monthly tasks as well as your short-term and long-term goals.

CHAPTER 3

How to Fit Research In with Everything Else

Time and Energy Management

> I get up every morning determined to both change the world and have one hell of a good time. Sometimes this makes planning my day difficult.
>
> —E. B. White

If you've just finished Chapters 1 and 2, you've been learning about the whats, whys, and hows of college, of research, and of changing something in the world through your discovery, creation, and analysis. What you may be wondering now is *when*: "When do I do all of this work?" and "When do I play?" This whole chapter is dedicated to just that question. It's about *time and energy management*, which means prioritizing what you have to do, from big goals to daily tasks, and planning when you'll get things done. You want to maximize your energy to change the world *and* have a good time when you are and aren't working. This chapter will help you figure out how long a task will take and then take that amount of time to get it done. We'll examine differences between high school time and college time, how to create and maintain a schedule, the time demands of underrepresented students, and of course, how to fit research in with everything else in college. In addition, as you learn to plan enough time for particular tasks, rather than rushing to complete tasks at the last minute, you will start to enjoy the actual working process. Even with a focus on the when, we won't leave the how and why behind. We also want you to examine the bigger picture: how you spend your time, why you spend your time in the ways that others suggest or prescribe, and how you choose to live your life. Becoming fully mindful of how you spend your time is an important part of your identity as a scholar and a researcher. You don't have to re-create the rat race that you may witness on your campus or in other research spaces. We want you to constantly ask yourself the question: *What am I doing it all for?*

WHY TIME AND ENERGY MANAGEMENT IS CRUCIAL TO THE RESEARCH PROCESS

Process is vital to researchers. Research isn't about just getting work done. In order to be successful at research, you need to make a serious commitment because of the many different tasks involved. Research cannot be done overnight, and for each project you will be spending more time on one task than another, depending on the scope of the work. For some projects you will need to analyze some form of text or material. For others you will need a large chunk of time for the creative process. Others will require time to collect data or conduct interviews.

In high school maybe you never thought about managing your time; maybe you went to class, did all your homework, participated in extracurricular activities, met your responsibilities at home, and had fun, without consciously scheduling your time. Or maybe you were overwhelmed by all you had to do, and looking back, you were never sure how you got it done. Or, maybe you were a great scheduler, diligently filling out and following your planner or calendar every day. Regardless of how time management went for you in high school (or if it happened at all) and regardless of how it's going now, this chapter is for you! That is, you can always manage your time better in order to increase productivity, be less stressed, and have more fun. Verrell and McCabe (2015) surveyed about 700 college students in order to learn about college readiness from the perspective of students. The results revealed time management to be the most frequently reported skill that students wished they had developed more in high school, and the most frequently reported skill that students wished to develop in college.

Even experienced researchers, like your professors and us, need a time-and-energy-management refresher every semester or so. That's because researchers are always doing something new and life is always changing. As a researcher, you will always be asking new questions and trying new ways to find answers. Each time, you will have to figure out how long those new tasks will take.

A review of the time-management literature (see Chapter 5 to learn more about literature reviews) showed that researchers who study time management have found positive impacts of goal-setting, prioritizing, and planning—all of which we'll address in this chapter (Claessens, van Eerde, Rutte, & Roe, 2007). The positive impacts include higher grades and decreased stress, as you will have more time for fun and for yourself. Most time-management researchers have learned about these effects by using specially developed surveys, but some researchers have analyzed activity diaries or conducted experiments or observations. Additionally, the literature review revealed that most time-management researchers examined college students in particular, so these findings are especially relevant to you!

ACTIVITY 3.1. TIME MANAGEMENT ASSESSMENT

Check the statements that apply to your current habits:

1. I am usually satisfied with what I get done each day.
2. I schedule time every week to read, write, and study.
3. I use my syllabi each semester to schedule all of my due dates.
4. I use a to-do list to write out my plans each day.
5. I use a calendar to schedule my to-do tasks by the day, week, month, and semester.
6. I schedule time for larger projects, papers, research, and exams well in advance.
7. I make sure to plan my time according to my goals for college and for after college.
8. I take advantage of the small chunks of time available during the day.
9. I can usually overcome tendencies to procrastinate.
10. I know how to minimize distractions when I need to.
11. I have enough time for all of my commitments.
12. I am ready to fit research into my schedule.

Source: Adapted in part from Petrie, Hankes, and Denson (2011)

Let's see how time management actually is going for you with the Time Management Assessment in Activity 3.1.

How did you do? What items are you currently good at? What items do you need to work on now that you're in college? It's okay if you have a lot to work on! This chapter will address all of these twelve items. Pick one or two items to work on for now, and keep those items in mind as you read the rest of the chapter. Try to return to the Time Management Assessment every semester or so to track your progress and check in with yourself periodically to make sure you are achieving your time-management goals. Time management takes practice—not a specific personality type—to develop new habits (Dartmouth College Academic Skills Center, 2001b).

HOW HIGH SCHOOL LIFE IS DIFFERENT FROM COLLEGE LIFE: PLANNING FOR COLLEGE LIFE AND BEYOND

Consider high school life in terms of time and energy. Take a few minutes to write down an answer to each of the questions in Activity 3.2 about your experiences in high school.

ACTIVITY 3.2: HIGH SCHOOL VERSUS COLLEGE

- Who schedules your time in high school?
- How do you know when it's time for class?
- Who makes sure you're following your schedule during the school day?
- Who makes sure you get your work done?
- Who makes sure you're where you need to be after school and on weekends?
- What happens when you don't follow your schedule—what are the repercussions?
- Whom do the repercussions come from?
- How do you feel about and respond to these repercussions?

It is likely that you answered that your teachers, school administrators, and maybe parents or guardians set your schedule, made sure you were getting your work done, and determined repercussions for your inability to follow a schedule. In college, teachers and school administrators are no longer responsible for your schedule. Your family may not be around while you're in college, or, if you continue to live with them while in college, they will be less informed about your day-to-day activities. You may begin to realize that no one will follow up on you if you don't show up for class or do your work, yet there is still work due. Also, if you are away from your family for college it'll take time for you to make that adjustment as well.

Look back at Activity 3.2. If you reconsider the questions as a college student, the answers will be more about you. In college, *you* do the scheduling, *you* make sure you follow your schedule, and it is *your* energy that is expended and *your* grades that suffer when you don't follow your schedule. For example, in high school, a teacher might take attendance, note who is not coming to class, speak with those students and maybe with their parents or guardians, and finally issue disciplinary action. In college, many professors do not take attendance, but students who do not come to class might miss important information, notes, and assignments, and consequently receive lower grades. A professor might actually take attendance and note who's not coming to class, but that professor may not speak directly to those students and, again, the students may receive lower grades. This is just one set of examples. As you have been learning, there is much more to college than just coming to class, but class is a first and very important step.

This transfer of responsibility from school faculty and staff to *you* as a student is emphasized in the "Top 10 Misconceptions Students Have about

College," written by Abela and Renfro (2002) when they worked in Student Support Services at Southeastern Louisiana University. Abela and Renfro emphasize that it is your responsibility to figure out what you need to learn in each course and across courses, to learn that material, and to make sure that you are fulfilling requirements for graduation.

In the following vignette, scholar Thomeka Watkins explains how her college schedule is different from her high school schedule:

> My overall daily schedule in college is much more compact than in high school, but my academic schedule is much looser. Because college students get to design their own schedule, I built mine my freshman year so that I had a balance between back-to-back classes and reasonable breaks between large blocks of classes. Having one class end at 10:50 am, for example, and not having my next class until 12 pm gave me ample time to have lunch, relax, organize myself, and stop by professors' office hours for extra help without having to rush to my next class. Classes are also not necessarily as long as in high school. While in high school I would have A and B block days lasting from 7:25 am to 2:10 pm with four 90-minute classes and a brief lunch break in between, in college I might have a 50-minute class in the morning starting at 10 am, then have classes from 12 pm to 1:50 pm and be done with classes for the day. During my longer days I might have class from 9:30–10:50 am and then class from 1–3:20 pm. Since I began college, my schedule has become also much more compact with club meetings, service activities, and fun events.

The Bigger Picture

Throughout this chapter, we emphasize the need to think about your bigger-picture goals as well and how to start expressing those goals to your research peers and mentors. Through our approach to time management, we are not advocating for you to necessarily re-create the schedules of current researchers. It is important to ask yourself: What are you doing it all for? What does each research endeavor mean in the larger scale? What about each research relationship? You may want to ask other questions: What were some of the drawbacks to the time sacrifices that some of your research mentors or models made? Would they do things differently if they had to do them over again? We ask questions in this chapter related to these themes in textboxes marked "Bigger Picture."

Even though you are taking on greater responsibility for your time once in college, that does not mean that no one is there to help you! We will help

Textbox 3.1: Whom to Contact at Your School

There are resources on your campus to help you with time and energy management. We encourage you to seek them out and to take advantage of them. You might start by asking your academic advisor where to go on your campus for time-management advice. In our experience, asking multiple people for time-and-energy-management suggestions can go a long way. Although not every suggestion will work for us as individuals, we often get a couple of tips from each source that we can use. For example, you will probably find one or more of the following departments on your campus:

- Dean of Students Office
- Student Support Office
- Student Learning Center
- Academic Skills Center
- Undergraduate Research Center/Program
- Library Research Desk/Office

You can probably set up a one-on-one appointment with someone at one of these centers or offices on your campus. Attending to the Activities in this chapter will prepare you to make the most out of your appointments. Know that these offices *want* you to use them as resources—that's why they exist! Don't be hesitant to set up these appointments as often as you need to. And of course, put them in your calendar immediately. For research scheduling, many campuses have offices dedicated to undergraduate research—another important resource.

Also seek out other offices that focus on energy management in particular, including perfectionism, stress, and wellness. At your campus, those departments might include Counseling Center/Services, Recreation Center/Services, and Dining Services.

you in this chapter, and we will also point you to resources that are available for you at your own college. Textbox 3.1 provides examples of college resources that may be available to you.

You can begin taking on greater responsibility by deciding what your goals and priorities are. Start with the tips and questions in Activity 3.3.

The time you spend planning will save you time later on. For example, steady, focused attention across the semester is more effective than cramming. Participating in research will help you develop these skills in steady, rather than crammed, learning.

ACTIVITY 3.3: HOW TO GET STARTED WITH COLLEGE TIME MANAGEMENT

Set goals and prioritize

Start with your major goals for the future and other goals for after college—the bigger picture! Use these long-term goals to set short-term goals and prioritize your activities for your time in college. For example, you may start to find out answers to questions like:

- What GPA do you wish to attain?
- What GPA do you need to accomplish your long-term goals?
- What courses do you need to prioritize?
- What kind of research experiences do you need to seek?

Plan

Use your goals and prioritized activities to make concrete plans for your time in college.

- Create a college plan: What goals will you focus on in each year of college? What courses do you need to take each year, including prerequisites for any research experiences you are interested in?
- Create a plan for the current year: What courses will you take each semester this year? How can you schedule your courses so that you are not taking too many credits at a time? What else will you accomplish each semester?
- Create a plan for the current semester: Use your syllabi to track due dates, which can then help you plan by the month, week, and day—we'll do this type of planning in the next section.

Source: Adapted from Vye, Scholljegerdes, and Welch, 2007

The Day-to-Day

You'll need a calendar to take care of the advanced planning required for research and college courses. A digital version is best for several reasons. First, it will allow you to set regular commitments that reoccur (e.g., weekly classes). Second, you will be able to easily move commitments when you need to, which may happen if a meeting or an exam gets rescheduled. Many calendars also help you plan for long-term goals. With a digital calendar, you also have flexibility: You can use it on a computer or mobile device or you can print it out if you want. More importantly, a digital version is much more difficult to lose or to have stolen than a paper version, especially if it's online. A digital version can also help you see the bigger picture of your day-to-day and month-to-month life. For example, Prof. Charity Hudley started using Franklin Covey Planners in 1994. She then started using Calendar for Apple in 2004, which she then synched with Google Calendar when it became available, and can look back to see her scheduled events over time. A lot of students and professors at William & Mary like to use Google Calendar, which is accessible from anywhere with Internet access, so you can use your calendar from your phone, your own computer, or a campus computer. You can also set up Google Calendar to sync to calendar apps, like Apple Calendar, on any device. If you already have a Google account, for Gmail, for example, you already have a Google Calendar—you just need to start using it!

Once you have a calendar ready to use, it's time to start scheduling. If you are just starting college, this can be exciting, because you will be able to schedule blocks of time on weekdays between 8 a.m. and 3 p.m. *on your own*—time that was previously scheduled for you. Plan your newly discovered time wisely! You'll be spending a lot less time in classes as compared to high school. For example, the average student at William & Mary takes 15 credits per semester, which translates to 4–5 classes, or 3–4 classes with labs and research. If you are taking 15 credits in a given semester, you will spend roughly 15 hours in your classes per week. That's not a lot of time compared to the approximately 30 hours per week you spent in the classroom in high school. This change means that in college you will have to do most of your learning outside of class, as suggested in the book *Orientation to College Learning*, written by Dianna Van Blerkom, a former professor and director in the Academic Support Center at the University of Pittsburgh, Johnstown. In other words, more time equals more responsibility. Van Blerkom also suggests that college students take advantage of their time during the day; you won't be able to get all of your assignments done at night. Activity 3.4 can help you get started creating your schedule.

We recognize that transitioning to a 4-year college from high school is a lot at once. For many students, a more gradual transition to college-level courses and then out of their previous living environment may be better so that as their academics intensify, their family and friend supports stay the same. When you add new friends and activities to the environment, a successful transition may take time. You are going to make mistakes, especially during your first year of college and your first time researching, as you learn to navigate academia, but resources are available to you so that you work smarter, not harder.

Activity 3.4: Creating Your Schedule

Required/regular activities: First, schedule all of your required and regular time commitments: classes, labs, outside work, meetings, eating, time with friends and family, etc. (Abela & Renfro, 2001).

Sleep: Schedule time to sleep. Sleep is essential to getting good grades in college. Sleep is also essential for researchers, because their job is to generate new ideas and questions, rather than to simply learn existing ones. Staying up late to work on your research presentation or data analysis or pulling an all-nighter cramming for that exam can make your brain work less effectively. We know that sleep is essential for academic success because there is lots of research on the topic (e.g., Eliasson, Letterieri, & Eliasson, 2010).

Study time: "How many hours do you estimate you should study for each course each week to do a job worthy of a research student?" (adapted from Dartmouth College Academic Skills Center, 2001b). To help you answer this question, try talking to your professors. Your college may have guidelines on how much time to spend outside of class per credit hour or per course. At the same time, each professor may have a slightly different expectation, particularly for students who will be researchers. On the first day, ask your professor how many outside-of-class hours per week or per class meeting you should expect to spend to be successful. Then schedule that number of study hours weekly for specific times on your calendar. Make sure to schedule extra hours when you have an exam or a paper coming up.

Energy: When do you have the most energy during the day? Biology researchers Horne and Ostberg (1976) developed a survey that assesses chronotype—whether someone is a morning person, evening person, or somewhere in between. Other researchers have used the survey to assess people from all ages and all walks of life and found that most people tend to perform better on a range of tasks involving critical thinking at their optimal time of day (e.g., Fabbri, Mencarelli, Adan, & Natale, 2013). Think about when you are at your best and try to maximize the work that you get done at those times. Take the morningness-eveningness survey yourself, found online at www.hhmi.org/biointeractive/are-you-morning-or-evening-person. You can even try to schedule your classes and research around your individual style. What should you work on during your optimal time? How will you use high-energy times for research?

Small chunks of time: When do you have small chunks of time (e.g., between classes) when you could get something done? For example, you might have a 1-hour break between your 1:00 p.m. and 3:00 p.m.

classes every Tuesday and Thursday. Take advantage of these times. Early in the semester, you can use those short chunks of time to plan for larger assignments, like research projects, and begin to jot ideas and to outline for those projects.

Breaks: Schedule breaks for relaxation, socialization, and leisure activities you enjoy. Breaks are important in their own right, and they also improve your focus when you return to work.

Be flexible: Leave open time each day for the unexpected, like new priorities or opportunities that might arise or challenges to your daily plan (Abela & Renfro, 2001). Research takes flexibility; an experiment may not go right the first time, a book or journal article may not be in your library and you'll have to request that your library get it on loan from another college, participants may not show up. Ensure that you make time for things you cannot control in research.

Review your time management: Schedule weekly time to review your time management and plan for the next week (Abela & Renfro, 2001). How is your schedule working for you? When and where are you getting your best work done? What might need changing? What time-management skills are you improving and what skills do you want to work on next? How much reading, writing, and other research work are you getting done at what times? Even though it may seem counterintuitive to spend your time working on time management, tweaking your skills in this area will pay off in the long run. It's also important to review your life goals and related priorities and how your current use of time reflects those goals and priorities.

Check your schedule daily: If you always have your phone with you, you can set calendar alerts to help you keep your schedule. If you're on your Smartphone or computer often, you can keep your calendar app open so your schedule is always right where you can see it.

Scholar Thomeka Watkins shares how she helped to smooth the time transition to college:

> Creating and maintaining a calendar starting early in my freshman year has made the transition tremendously easier. With a syllabus and assignments to keep track of for every class, it is not difficult for due dates to sneak up on you. With my calendar, however, I can easily keep track of the dates of every major assessment, project due date, and paper due date for every class and plan accordingly. It also helped me to know how much of my weekend I needed to dedicate to preparing for the upcoming week to avoid falling behind.

Scholar Christine Fulgham describes how planning and scheduling has improved her college experience:

> Calendars and planners save my life. Without my planner I am always scared I'm going to miss something. I use Google Calendar to keep track of my classes and rarely miss meetings or events. I keep all my meetings, events, assignment due dates, and appointments in my paper planner so that if my phone or computer dies, I have all my important information with me. You'd think that after the first few weeks of class you would remember all your classes, right? Not necessarily. I kept forgetting about my Theatre Photography class on Fridays, so I almost missed it every week. I left my planner at home when I returned from Spring break. I had to have my mom mail it to me the next day because I realized I relied heavily on my planner to keep track of what day and what time all my events happened. I sort of wandered around in a confused daze hoping to not miss anything. When the package arrived, I can honestly say that I have never been so thankful to see a group of bound papers in my life.

ASSESSING YOUR SUCCESS: BEING RESEARCH READY

How to Fit in the Demands of Research

We've been reiterating throughout this chapter that research takes advanced planning and multiple steps, so let's take a look at some of the steps involved and how to schedule them. Research takes collaboration, lab time or other data collection and analysis, and creative and rehearsal time, in addition to extra reading and writing. You'll need to collaborate with a professor to work on research. One of the best ways to communicate with professors, whether you are hoping to work with them or already are working with them, is to make use of their scheduled office hours. Office hours are times that professors have dedicated to talking to students—there is no need to make an appointment, as you can expect your professors to be there each week. To ensure that you make office-hour visits, put them in your calendar. Treat office hours, especially with a research mentor or potential research mentor, as fixed commitments. Fixed commitments are things that are scheduled like your classes: Put them on your calendar, don't schedule anything else for that time, and schedule the time before and after to make sure you go. The same is true for lab time, reading, and writing: Treat them as fixed commitments that are prioritized the same way that class time is prioritized. You might be thinking that class + appointments + office hours + lab + read-

Textbox 3.2: Bigger Picture

What do you want to learn, what do you want to gain, and how do you hope to grow as a person? What are you willing to sacrifice? What do you want to be careful that you don't lose?

ing + writing = a lot of fixed commitments! Know that college academics (including research) are a 40-hour-a-week job. If you sleep 8 hours a night, that still gives you 72 hours a week for relaxation, family, social activities, outside work, any challenges that may arise, and extra work during your busier weeks. In Textbox 3.2 we provide some questions to help you focus on the bigger pictures of time management that research makes possible.

So with all of the students that your professors have, how do you stand out in class and make sure the professor knows who you are and can see your potential as a researcher and scholar? In middle and high school, many of your teachers probably took the initiative to get to know you and to make sure you reached your academic potential. As with other responsibilities in college, making sure you know your professors and your professors know you is now *your* responsibility. Scholar Kristin Hopkins explains why this responsibility is important:

> College curriculum is not as personal [as high school] unless you make it personal. If it's a class that I know I'm going to have, not even a lot of difficulty with, but just a little, I'll make sure that the professor knows my name and I will email them with questions or go visit them during office hours. I just have to make that effort if I feel like I need to. Beyond getting extra clarification, office hours have allowed me to get to know my professors on a level that goes beyond the classroom. I'm learning that it is critical to make personal connections. Professors definitely notice who is making the extra efforts and who is not.

When you schedule your extra reading time for research, it's important to know that you will spend much of that time searching for sources you need to read. When you work on research, your professor will not necessarily tell you what books and articles you need to read, but you will be figuring that out on your own (there's the theme of responsibility again). There are librarians at your college library who can help make this search process as efficient as possible, so you may want to schedule at least some of your fixed reading time at the library. Even though most of your time finding resources will be spent online using databases or other resources (see Chapter 5), having a librarian help you identify these resources and show you how to use them in the most efficient way possible can be a real time-saver. It's also a good idea to work on skimming skills so that you can efficiently determine what sources will be most important to inform your research. To learn more about reading for research, check out the books *The Craft of Research* (Booth, Colomb, & Williams, 2008) and *Writing Your Journal Article in 12 Weeks* (Belcher, 2009). The authors of both books are professors who are researchers.

Writing is an important part of the research process. In interviews we've conducted, college professors indicated that they want students to know the importance of revision and multiple drafts (Franz, 2016). They also want students to see writing a paper as a process rather than as a task to be completed the night before the due date. In Chapter 5, you will learn more about writing in college and in research, including the steps you need to take. All of this information is important for time management, because if you see writing as an ongoing process, you will schedule it more regularly. For example, if you have a paper due in the coming weeks, ask yourself the questions in Activity 3.5.

In the summer you may find yourself with large chunks of time that weren't available to you during the school year. The summer is a great time to work on research, to refocus on your life goals and long-term priorities, and to make sure you take some time off. You can make research your summer job by working in a professor's research lab or receiving a summer research grant, both of which pay you for your research time. Professor availability may vary in the summer, so it's important to check in advance to see if professors will be around or more intensively focused on research in ways that they are not during the academic year. Even if you can't find existing research grants for undergraduates at your own college, you can do research at other universities or research centers around the country, as we explained in Chapter 2. The National Science Foundation funds Research Experiences for Undergraduates (REU) at sites across the country. An REU usually includes a small salary, housing, and travel expenses. You can find out how to apply to work on a particular REU research project online at www.nsf.gov/crssprgm/reu/index.jsp.

Also take time each summer to plan and review your long-term goals. What are you planning to change in the world? How can you make that change? What will you do after graduation? How can you make the most out of college and your summers to reach those goals? Summer may be a good time to meet with a professor about some of your goals, but know that professors take vacation time too, so you should plan a time to meet in advance.

Use the summer to plan your own time off too. It's important to recharge in the summer so that you are ready to go in the fall.

Activity 3.5: Planning for Research Writing: Start from the End and Work Backward

Smart planning is essential for success in writing. Psychological research has shown time and again that people tend to underestimate the time required for assignments and tasks—a concept known as the planning fallacy (Kahneman & Tversky, 1979). Buehler, Griffin, and Ross (1994) tested hypotheses about the planning fallacy in a series of studies involving both actual and researcher-given assignments. When the researchers gave undergraduate students assignments to do, such as a computer tutorial, or asked students to report on actual assignments like course papers or honors theses, students tended to leave the work until the last minute and underestimate completion times for the assignments. The students participating in the studies failed to accurately recall previous times when they had similar tasks and thus forgot that unexpected obstacles such as sickness or helping a friend deal with a crisis can occur and delay things. The good news is that other researchers have found, through time-management surveys along with task-completion experiments, that people who see themselves as good time managers were better able to estimate the time to complete tasks (Francis-Smythe & Robertson, 1999). You are now armed with this knowledge, and thus should ensure that you schedule even *more* time than you think it will take. Schedule just a few steps each week until the paper is due. If you start from the last step and work backward, you can make sure you schedule enough time and can prevent yourself from cramming your writing in at the last minute.

- When will you proofread your research writing for the final time?
- When will you revise your drafts?
- When will you draft the second half of your writing?
- When will you draft the first half of your writing?
- When will you outline your writing?
- When will you read and take notes on the sources you need?
- When will you find the sources you need?

If you work well with a detailed plan or are simply ready for the next step in your scheduled writing, read the books *How to Write A Lot* (Silvia, 2007) or *Writing Your Journal Article in 12 Weeks* (Belcher, 2009), written by professors who are researchers.

Procrastination, Perfectionism, and Balancing Your Energy

Sometimes the reason we don't stick to our schedule is not just because we didn't schedule things for the right time or right place, but because we frame our tasks in ways that prevent us from completing them, and we put our energy into unrealistic goals and work avoidance, rather than the immediate task that we need to complete. We may do so often without realizing it, out of fear, perfectionism, and/or habits of procrastination. A survey of 342 college students found that between 30% and 50% reported a high level of procrastination on academic tasks such as writing a paper or studying, as assessed by a validated questionnaire. As a high-achieving student, you probably have faced at least one of these obstacles to time and energy management. Two separate sets of researchers, Rothblum, Solomon, and Murakami (1986) and Akinsola, Tella, and Tella (2007), wanted to find out how procrastination is related to academic performance for college students. Both sets of researchers surveyed students using a procrastination scale. The results suggest that procrastination is extremely common among undergraduates and even more so among those who are high-achieving scholars. This research also shows that students who procrastinate more do worse in their classes and have more anxiety than those who procrastinate less. Eastern Illinois University provides questions you can ask yourself about procrastination, which can be found online at www.eiu.edu/~counsctr/Procrastination.php. The obstacles described may be particularly prevalent in research that does not have fixed deadlines. See which items you might need to work on.

Once you have an idea of what you need to work on from the questions from Eastern Illinois University, set up an appointment at the office at your school that helps students with time and energy management (see Textbox 3.1: Whom to Contact at Your School). You can also make an appointment at your college's counseling center or services, another resource that you should take advantage of during college. Counselors are trained to work with you on energy management and internal obstacles to getting tasks done. For any such appointment, have a few questions ready that pertain to the particular areas in which you'd like to improve. For example, you may ask, "What are some strategies for preventing _____?" After your appointment, make a follow-up appointment for the next week or so. Your counselor will want to know how things are working out for you, in order to help you achieve your goals.

Procrastination can be related to perfectionism, which is the tendency to hold unrealistically high standards for yourself, be overly critical of yourself, and be excessively concerned about making mistakes. Researchers estimate that about one-fifth to one-third of college students are perfectionists (Boone, Soenens, Braet, & Goossens, 2010). Psychologists have developed self-report measures where you can assess the degree to which you exhibit perfectionism, which some have determined is a personality trait. One such scale, the Almost Perfect Scale—Revised (Slaney, Mobley, Trippi, Ashby, & Johnson, 1996) was modified by psychology professors at Georgia State University and the University of Florida (Rice, Richardson, & Tueller, 2013). Here are a few items from their Short Form of the Revised Almost Perfect Scale to help you determine whether you may have perfectionistic tendencies, which may spill over into your research activities:

> "I have high standards for my performance at school."
> "I expect the best from myself."
> "I have a strong need to strive for excellence."
> "I am never satisfied with my accomplishments."
> "I rarely live up to my high standards." (p. 372)

If you identify with any of these statements, you might be a perfectionist. Although perfectionism can often yield success in school and other areas of your life, it can lead to feelings of inadequacy. When it comes to conducting research, the tendency to be a perfectionist may lead to especially negative feelings and may affect your success. Overscheduling—thinking you can handle research in a limited amount of time—is also a challenge because when you can't, that leads you to retract from trying and then to procrastinate and to think, "What's the point, I've tried?" Research is all about questions that lead to more questions, and no research is perfect or finished. For example, if you are conducting a study on perfectionism at your college and you learn that 50% of your peers have perfectionistic tendencies, the next logical research question you might ask as is "Why?" This can lead you to want to conduct additional research investigating this question until there is no end. When you learn more about the underlying factors of perfectionism, you may then want to learn about preventing perfectionism. When you are overly motivated to do everything, it can also be a challenge to know when what you've done is enough so that you don't spend too much time on research or other assignments.

As you can see, being a perfectionist may lead you to be unsatisfied with the open-endedness of research such that you will never find *the* answer. As a researcher, it's good to know that no research project is ever perfect. You can never design the perfect experiment or have the perfect performance. You can never write the perfect paper that has no mistakes. In fact, many research articles, especially in the social and natural sciences, end with a "Limitations" and/or "Future Directions" section. These concluding sections acknowledge that improvement, not perfection, is a goal for researchers. Research is a continual learning process, in which you analyze possible limitations of your work and make plans to answer future questions. In short, you can't do everything as an undergraduate student or

TEXTBOX 3.3: BIGGER PICTURE

What are your goals for your actions, and how are you spending your time? Can you use your research skills to discern if what you are doing is *really* worth your time?

even as a professor—research is a process that occurs across many lifetimes. The questions in Textbox 3.3 are designed to help you reflect on the bigger picture questions about how you spend your time so that you can prioritize in the face of perfectionist tendencies.

So give yourself a break and make sure you take credit for your accomplishments, big and small. Set realistic goals and realize that it is impossible to be perfect. Think about what the worst possible outcome could be if you are not perfect on a particular research project or class assignment—this can help put things in perspective for you—and learn from the mistakes that you make. Researchers also handle any tendency toward perfectionism by addressing the audience of their work. When you find yourself wanting your work to be perfect, ask yourself, "Perfect for what audience? Who will be, and who should be, reading and hearing about my work? What will they want to know?" Focus your energy on communicating to that external audience rather than on satisfying your internal perfectionism. And remember, strong research is *difficult* even for the most seasoned scholar.

So that your research doesn't suffer, use your counseling center as a resource for working on perfectionist tendencies common to high-achieving students.

According to Vye, Scholljegerdes, and Welch (2007)—two psychology professors and a career services director—procrastination and perfectionism are also tied to anxiety and poor health. Notice that perfectionism may mean you are putting your energy into fear of negative outcomes, "shoulds," and unrealistic beliefs about others' achievements. If you ever spend energy on any of these thoughts or worries, imagine what could happen if you rechanneled that energy instead into your own health. When students procrastinate or face other obstacles to time and energy management, their health, and especially their sleep, is often the first thing to suffer. But when you get a full night's sleep (that's 8–9 hours, not 6!), you will have the energy you need to work on your tasks efficiently and be less prone to distraction. Textbox 3.4 takes a bigger picture view in helping you address obstacles that lead to procrastination.

The reality for many students is that their schedules are not entirely their own. Some students have responsibilities—families, jobs, and other

TEXTBOX 3.4: BIGGER PICTURE

Where do you want your life (not just your research or your academics) to lead?

things that are important and enrich the research perspectives we need to have for the future. Others have health needs and physical and mental differences that require more time than our society (or especially academic culture) seems to afford. It is important for student and faculty researchers to work together to create a world where people of different paces of accomplishment are welcomed and valued. You can start by talking to faculty about such needs before committing to research with them and finding faculty when possible who are inclusive and equitable. It is also a good idea to ask your fellow students and professors you respect to give you advice about which research mentors may be more supportive than others. When a student approaches us, we often speak to the student about their preferences for not only the research area but also the personal style they would like from a research advisor (e.g., hands-on versus more independent), so the student will be matched up with a mentor who will be supportive and respectful of their personal needs.

Academics. In college, *academics* are a full-time job. That means at least 40 hours a week spent on your courses and research. Remember that class is an important first step. If you don't do well in classes, it's hard to do well in research for a few reasons. First, your courses provide much of the knowledge and skills you need to excel in research. Second, many research opportunities require particular GPAs or prerequisites. The reality is that, based on your preparedness and your own high school experience, academics may take more or less than 40 hours a week, so the amount of time varies for different students. It's not fair but it's something we all have to live with. If you take extra time, that's normal, and if you take less, help someone else on their journey. The next sections are designed to help you with the specifics of balancing academics with the rest of your life while in college.

Research. To make the most out of your classes, as you have been reading, participate in *research*! Research takes your highest levels of creativity and focus, so make sure to prioritize research work as part of your 40 hours.

Work. Of course, you may need to work to support yourself financially. So that your *paid work* doesn't conflict too much with your 40 hours of academics, get paid for your research! Get a job in a professor's research lab. Apply for research grants. You could even consider getting a job at your college's library or writing center to strengthen your research and writing skills.

Extra- and Co-curricular Activities. Graduate school admissions are primarily concerned with your undergraduate GPA, faculty recommendations, and, for some programs, admission test scores. Unlike undergraduate admissions, graduate school admissions do not tend to focus on your extracurricular activities. Graduate schools want to know about your potential to do well academically in their program, not about your potential to be well-rounded. Researchers who study graduate school admissions examine how well certain factors can predict student success in graduate programs. For example, Dodge and Derwin (2008) examined differences in graduate GPA and earned graduate credits between students who were admitted based on rubric-scored portfolios and students who were admitted based on more traditional measures like undergraduate GPA and admission test scores.

If part of your relaxation and health activities include participation in a sport or in music or another art as a hobby, just remember that you don't need to—and shouldn't—do it all! The section below on Saying No talks about how you can fit one of these activities, whichever is the one for you, into your schedule.

After reading the above list, you might be worried that you will have to let go of some of your interests in college. If you participate in research, that's not the case! Bring your interests and maybe your former extracurricular activities, whatever they may be, into your research. You can research music from a linguistic lens, video games from a sociological lens, sports from a historical lens, food from a psychological lens—these represent just the tiniest fraction of examples to give you an idea of the endless possibilities!

What about community service? You probably participated in community service in high school. Maybe you enjoyed it, maybe you had to do it, maybe you learned from it, maybe it's an important part of who you are. Guess what? There is a whole genre of research dedicated to the importance of community engagement. It's called community-based research. You can do community-based research in any discipline. You might also be able to take service-learning courses at your college, which connect community service with academics. Again, these classes could be in any subject. We talk more about community-based research in Chapter 7.

Because of your 40-hour academic week and your need for work-life balance, you may have to say no to some potential time commitments. That is, friends might ask you to join clubs, sororities or fraternities, and other organizations; professors or offices on campus might ask you to volunteer for various activities. Keep in mind your long-term goals when you decide what to agree to and what to say no to. In college (and later in life), you will probably more often have the opportunity to say no than you had in high school. A key to saying no comfortably is to remember that it may be what is best for you. It's okay to *say no* to new time-consuming commitments or to quit something you don't have time for anymore. Or, if you do decide to commit to an activity or organization, you don't have to do everything. It's okay to delegate responsibilities and tasks to others. Just because you think a task might be done better if you do it doesn't mean that you shouldn't give others the work to allow yourself more time for other things.

Saying No

High-achieving high school students on the path to becoming high-achieving college students often need to distinguish between what it means to be high-achieving in high school and what it means to be high-achieving in college. In high school, many students participate in multiple extracurricular activities—perhaps intentionally to make themselves better candidates on their college applications. By extracurricular we mean structured activities that are not part of academics, such as clubs, sports, and volunteering in the community. In contrast, in order to reach their maximum achievement potential in college, high-achieving students learn that in fact they can't and shouldn't excel or even participate in everything. We next describe how you might need to divide and prioritize your time, explaining why you shouldn't participate in everything.

Saying No Socially. Students often feel like they need to be excelling not only at academic-related activities, but also social relationships. While being a good friend and family member is very important, some social activities can take away from the reason you are in college: to reach your highest potential! With that goal in mind, it's okay to tell others that you don't have time when you absolutely have to get work done; to *NOT* answer phone calls, emails, and texts just because they show up, and instead to schedule a time to answer email and return phone calls; to block distractions like Facebook and Twitter to get stuff done; and to not multi-task—it's not for everyone! Try mono-tasking instead! Also, try setting a timer to help you block distractions and avoid the temptation to answer every request for communication immediately.

How to Postpone Requests for Your Time. It is probably easier said than done to change habits of frequently checking and responding to calls, emails, texts, Facebook, Twitter, Instagram, Snapchat, and other forms of communication. There is no wonder checking messages and updates is such a time commitment: There are so many ways for you to get messages and updates in the first place. Try to gradually change your habits. Use a timer and start with small blocks of time for work and frequent times for all of those communication venues. Then, you can work up to longer blocks of time for work and less frequent times for communication. For example, start with 10 or even 5 minutes on your timer. Silence your phone and place it out of reach. On your computer, have open only the applications and windows you need to get your work done. When the timer goes off, set it for 5 minutes to check your phone, email, etc. Keep repeating this cycle until you get good at blocking your phone and social media during timed work. Then try to work for 15 or 20 minutes between breaks, and later on maybe 25 or 30. And remember to take longer breaks or a nap when you need to. Try timing your longer breaks as well.

How to Say No. So you want to say no to a new time commitment, but how should you phrase it? Kerry Ann Rockquemore and Tracey Laszloffy (2008) have many suggestions on how to say no in their book *The Black Academic's Guide to Winning Tenure*. Rockquemore is a professor and expert in faculty development, especially for scholars from underrepresented backgrounds. Laszloffy is a therapist and faculty coach. Here are some of their suggestions for saying "no":

- "That sounds like a really great opportunity, but I just can't take on any additional commitments at this time."
- "I am in the middle of ________, ___________, and __________ [fill in the blanks with your most status-enhancing and high-profile service commitments] and . . . I'm unable to take on any additional service."
- "I'm not the best person for this, why don't you ask _____________."
- "If you can find a way to eliminate one of my existing service obligations, I will consider your request."
- "No" [look the asker in the eye and sit in silence]. (Rockquemore & Laszloffy, 2008, p. 118)

Rockquemore and Laszloffy remind us that people "accept 'no' for an answer" more easily than we might expect! But we also suggest you prepare for follow-up questions, plan out your answers with the support of advisors and mentors, and be confident in your answer to our driving question: What are you doing it all for?

THE EXTRA DEMANDS ON UNDERREPRESENTED STUDENTS: WHEN YOUR SCHEDULE ISN'T ENTIRELY YOUR OWN

Underrepresented Scholars

If you are a student from a background that is underrepresented on your campus or in higher education in general, chances are you have extra demands on your time. If you have a family or have to work to support yourself and others, your schedule may be further limited with additional unexpected events. Students may find that they spend extra time dealing with experiences, events, and situations related to their race and gender (see Chapter 6 for more information). In addition, professors and campus administrators might be especially interested in your perspectives. You have a lot of insight to add to research and to what's going on on your campus. You may find it important to make sure your voice is heard and your perspectives are represented in research areas and academic disciplines. If so, incorporate these efforts into your own research and academic workweek. It will also be necessary for you to say no at times, especially to requests that won't benefit you in the long term. Remember your big goals! And be sure to take the advice in the "Saying No" section above.

Student Athletes

Student athletes have additional demands on their time. Although participating in a sport can be rewarding in many different ways, keeping up with everything during a season with regularly scheduled practices and competitions can be difficult. Coaches may not understand that you need to miss a practice because of a lab or study session, and professors may not be supportive of your athletic schedule if it involves missing classes and exams because your team is traveling across the state or the country. To try to limit the negative effects that a sports schedule can have on your academic performance, give your professors your game/meet schedule as soon as you have it, which is usually at the beginning of a season. Most athletic departments will provide your professors with official documents with your schedule, but you should talk to your professors during their office hours about their policy for missing classes, labs, and exams. Some faculty may be able to meet with you once you get back to go over questions you have about what you missed, but for other classes you might need to befriend someone in the class who is willing to share notes from the classes you missed. Remind your professors the week before you miss a class as well as the morning of the class so that they remember where you are. They will appreciate the reminders. Finally, make use of all of your time! During long bus trips to away games during the soccer season, Prof. Dickter, when she was a student-athlete, would put on her headphones and do data entry for her psychology lab. She also had other teammates read over and edit her papers on the bus, particularly those who were upperclassmen in the same major.

Student Musicians/Performers

Music and other art performance schedules are often fixed. When the baton drops or the curtain rises, that's it. So student musicians and performers have to schedule rehearsal and performance as fixed time and ensure that they are ready to perform. Notify faculty and research supervisors about your rehearsal and performance schedule. Prof. Charity Hudley was a singer in college and found that her professors were very supportive of her schedule and would modify course meeting times and deadlines to accommodate performances. They would even come and cheer her on. In return, the skills that she gained as a musician transferred into her research (her honors/masters thesis was about the language and music of Bessie Smith) and gave her the ability to manage multiple fixed commitments.

Students Who Must Work

It can be difficult to maintain a job throughout school while trying to sustain good grades and to get the most out of your college experience. Finding a paid position or internship that is relevant to your intended major or career is ideal, in that it looks great on your resume and can give you experience within a field. Work-study jobs, which can be set up while you are working out your financial aid plan with the college, are often good opportunities in that they offer flexible hours and decent pay. If you are working an off-campus job, it is absolutely crucial to plan ahead when you are working. Take your syllabi at the beginning of each semester to your employer and try to plan your weekly work schedule accordingly, so you can ensure that you are working less when you have papers due and exams to study for. You can also try to find a job where you work during times when you are not at your peak time of day. For example, if you focus on academics better in the morning, find a job where you work at night. Prof. Dickter worked part-time as a waitress in college and found that knowing she needed to work the dinner shift forced her to make better use of her time during the day. Instead of hanging out and doing nothing for the 2-hour break she had between morning classes, she would study so she could go to bed after her shift was over. You can also be an entrepreneur. Prof. Dickter approached one of her professors, who liked hosting dinner parties with other faculty, with the idea of helping clean up after the dinner parties. She earned some money for a few hours of washing dishes and cleaning the kitchen, and her professor appreciated being able to enjoy the parties and not worry about cleaning. It was a great opportunity for Prof. Dickter to get to know some professors in different departments (and have some great leftovers).

When Challenge Strikes: Acute Interruptions

"How often is your daily plan destroyed by urgent interruptions?" (Dartmouth College Academic Skills Center, 2001a). Sometimes these interruptions, maybe classified as emergencies, make everything else you have to do seem unimportant or impossible. How can you cope with these challenges and stay on track for your big goals?

First, remember to prioritize your health. If you sacrifice what you need to keep yourself healthy, you will not be able to effectively keep up with your challenges or your day-to-day responsibilities, especially over the long term.

Addressing extenuating challenges may mean that you have to ask your professors for extensions on research or class work and you may not be able to plan in advance. For example, if a family member suddenly gets sick, it's understandable that you would not be able to give your professors advance notice that you would miss class time or lab time. That's okay too. Although the last thing you want to deal with when going through a challenging situation is your academic life, it is important to let the faculty know as soon as you know what the circumstances are, how long this will likely affect your attendance, and any other details you have. It is not necessary for you to go into detail about the exact circumstances, although you can talk about the situation with them if you want. It's very important to communicate with everyone with whom you work: professors, your research advisor, students you're doing research with, and others. It's okay to send a short email from your phone rather than a formal email while you deal with your personal situation. In cases where you are going to miss more than a couple of classes or where you want to ensure that your professors do not penalize you for absences that are beyond your control, you can report your circumstances to the Dean of Students office or another office on campus that serves students' personal interests. These offices will typically contact your professors and inform them that they need to take your situation seriously and give you extensions as necessary. Everyone deals with challenging situations like these differently, and you should cope with the circumstances in the way that works best for you. In suffering a family loss, for example, some students do better when they come back immediately to school and focus again on classes and research, while other students need more time with their family and do better putting off assignments and classes. You need to decide what is best for you and, as long as you communicate with faculty and your institution, you can handle the situation in your way. In some cases, such as a personal or family illness or medical emergency, you can withdraw from the semester, which allows you to leave school and make up the work at a later time. If working through the challenge by staying enrolled in school for the semester may hurt your mental or physical health or your academic performance, withdrawing might be a good option to consider.

CONCLUSION

You are now ready to turn your attention to the mentor relationship. Such relationships take time but are worth it because they lead to a lifetime of learning. Now with a better sense of why you want to do research, what research is, and the time that you have to devote to research, your research relationships will be that much stronger!

CHAPTER 4

Research with Professors and Mentors

As we have introduced you to college research, we have pointed out that you do not have to undertake this transition alone; there are many people and resources that you can work with on your campus and beyond. Your professors will likely be your most important resources and collaborators as you become a college researcher. In fact, positive relationships with professors are likely to have the longest-lasting impact on your personal growth and academic development (Malachowski, 1996). The Gallup organization recently polled 30,000 college graduates to see what academic factors contributed most to their happiness in their work. They found the most important things were: (1) a professor who cared about them as a person, (2) a professor who got them excited about learning, and (3) a professor who encouraged them to pursue their dreams (Busteed, 2015). From your professors you can learn the processes of conducting, presenting, and publishing research. Professors are also very important for helping you achieve your goals for the future through network connections and recommendation letters.

In many college courses, you are assessed only a few times during the semester. For example, in some classes you will have only two to three exams a semester or a few smaller assignments and one big final paper. Because your grades in your courses are based on just a few chances for evaluation, it is imperative that you see your professors early and often! In some classes you might need to meet with a teaching assistant (TA) rather than the professor, particularly if the class is large. The TA is likely a graduate student in the department and in many cases is at least partially responsible for grading exams and papers. In most cases, the TA will need to dedicate a specific amount of time to help students in your class, so take advantage of your teaching assistant's time and knowledge.

Several research studies have investigated the aspects and benefits of mentorship for undergraduate students from underrepresented backgrounds. For example, Smith (2013), a sociology professor, interviewed undergraduate students and their mentors. She created a model for mentoring undergraduate students that focuses on unveiling the hidden curriculum, or the often-untaught skills, strategies, and knowledge necessary for success in college and after. Smith found that when students established relationships with professors, especially early in their college careers, they were able to maximize their opportunities for graduate school. This finding has been replicated by other researchers as well, such as Saddler (2010), an education professor, who explored mentorship in summer research experiences specifically.

In this chapter you will learn more about how to approach professors by email and during office hours. You will also learn how to build positive relationships with your professors and find research mentors who will assist you with research opportunities, offer individual support in developing academic and professional skills, and write strong recommendations for you as you pursue graduate school and careers after college. We encourage you to visit your professors during their office hours and ask them some of the questions that we pose throughout this chapter and questions that you think of while reading this chapter. Read Textbox 4.1 to understand key terms we use throughout this chapter.

NAVIGATING THE ACADEMY: WHAT IS A PROFESSOR?

So what exactly is the job of a professor? What do they do? How are professors different from high school teachers? Ask your professors for their take on the answers! Although on the surface it might seem that the teachers you had in high school and the professors at your college do similar things throughout the day, the two roles often comprise different responsibilities. One big difference is that professors often divide their time between teaching, researching and writing, advising students and often other faculty, service (including serving on committees at their colleges and serving their academic areas and the community), and outreach. How much time they devote to each job area largely depends on their position and their school. When you find a professor who can serve as a research mentor, you will not only learn about research, but also about what it's like to be a professor and even how you could become one. When you understand the jobs of professors and the structure of the university, you can navigate your college more successfully as an informed scholar. In particular, the information below can help you decide whom you'd like to consider as a faculty research mentor.

If you've already looked up some of your professors on your college's website, you may have noticed that there are different titles for faculty members; these titles are explained in Table 4.1. Some of the faculty who teach your classes are *tenure eligible* (also called *tenure track*) and others might

Textbox 4.1: Key Terms

Advisor: usually a professor who acts in an official or assigned role. You will probably call the professor who supervises your research project or honors thesis an advisor. Although an advisor is an official role, you will probably have to seek out a research advisor on your own. This chapter will help you with that process.

Mentor/Research Mentor: usually a professor who acts in a less official role than an advisor, but note that an advisor is often also a mentor. You might consider any professor who gives you sustained guidance in academics or research a mentor or research mentor. As with an advisor, you will probably need to seek out research mentors on your own. The section "What is a Mentor?" will explore the role of a research mentor more closely. In this chapter, we realize that you will be working with both research mentors and advisors, and we use the terms interchangeably.

Faculty: usually a college/university employee with an academics-related appointment, including teaching, research, and/or academic administrative positions. We will look at different academic positions in the next section.

Professor: sometimes synonymous with *faculty*, and sometimes referring to a more specific position or set of positions. We'll define this further throughout the chapter.

Table 4.1. Titles of Faculty at American Universities

Tenure-Eligible Faculty	Non-Tenure-Eligible Faculty
Assistant Professor: usually 6 years	Adjunct Professor: teaches courses with no research responsibility
Associate Professor: tenured	Visiting Assistant/Associate Professor: teaches courses and may do some research
Full Professor: highest promotion	Clinical/Of the Practice Professor: provides practical instruction/supervision and application of practical knowledge
Department Chair: heads an academic department	Instructor/Lecturer: teaches courses
Emeritus Faculty: retired professor who still contributes to the college	
Dean: presides over a specific academic unit (e.g., Arts and Sciences)	
Provost: senior academic administrator	
President: presides over entire university	

be *non-tenure eligible* (also called *non-tenure track*). Just as you are working toward graduation and graduate school, your professors are working toward the next step in their careers. When tenure-eligible faculty are hired, they are referred to as assistant professors and begin a probationary period (typically 6 years), which is followed by an evaluation. Earning tenure means that a professor has passed through the probationary period and is then called an associate professor. Earning tenure provides increased job stability. Generally, tenured professors cannot be fired without due process. Increased job stability can have several benefits, including flexibility in research and teaching. For example, faculty who receive tenure can start investigating a new research area or develop a new class without risking a major career setback if the research or class is unsuccessful. Say you have a research idea that a tenured professor isn't necessarily an expert in; the professor may be willing to learn more about that area and spend time supporting your research project because of the job security associated with tenure. Tenured professors may also do research in more risky or controversial areas because tenure ensures job security and protection. Several years after earning tenure, associate professors are evaluated for promotion to professor, often distinguished from other types of professors with the title full professor.

Evaluation for tenure and promotion often includes review by administrators and peers at the college and/or at other colleges. Faculty are often evaluated on three tiers: research, teaching, and service. For research, professors need to engage in scholarly activities such as conducting research and publishing in peer-reviewed journals or books, as well as presenting research at conferences. The tenure process is usually 6 years long, in part because publishing research can take a long time. Faculty may be judged on the research quantity, research quality, and the prestige of publication and presentation venue (e.g., reputation of the publishing company, academic journal, or conference). For teaching, professors are evaluated on the type and number of classes they teach and the quality of their teaching as evaluated by other professors and by students. For service, professors are expected to serve on committees at department, college, and national levels. Although most colleges use this three-tier system, the degree to which research, teaching, and service is valued differs depending on the goals and size of the college. Other colleges may also branch off from the standard system and include evaluations based on outreach and work on diversity and inclusion. Understanding how your professors spend their time and how the college evaluates its employees will help you understand and navigate your college.

In addition to the titles professors can earn, there are also administrative positions that they can apply for. For example, a professor, usually after becoming a full professor, might become a chair of a department, which is a leadership position within a given department. A professor might also become a dean of the whole college (if the college is small) or of a group of faculty (e.g., science departments) or of a school within the college (e.g., School of Education or Business, if the college is larger). For these positions, faculty focus on overseeing a group of professors and working with different departments and schools across the campus. Although these positions can be prestigious and sometimes provide opportunities to improve the college, professors in these positions typically have less time for teaching and research because of their additional administrative responsibilities. If you ask your professors, they will likely be forthcoming with you as to why they chose their administrative position or why they didn't become an administrator. If you come across an emeritus professor, that's a tenured professor who is retired and has been approved to keep certain benefits from the college such as office space or access to library or other campus resources.

In college you might have classes taught by graduate students, teaching assistants, or non-tenure-eligible faculty, who may have titles such as instructor, lecturer or senior lecturer, visiting professor (hired for a temporary time period), affiliated faculty/professor, executive professor, professor of the practice, or clinical faculty/professor. Many of these titles are used differently at different colleges, but most of them often indicate that teaching and/or directing is the primary responsibility of the professor. Non-tenure-eligible faculty typically do not have many research responsibilities, although many of them have research experience and advanced degrees that require research and therefore they are often excellent research mentors. The last two titles, however, professor of the practice and clinical professor, are professors who are hired to teach in areas such as business, education, or medicine based on their professional experience, and so they do not necessarily hold advanced degrees. Non-tenure-track faculty may work full-time or part-time; part-time faculty are often called adjunct professors.

The role of mentorship and research in tenure/promotion evaluations varies across colleges and departments. Some schools and departments highly value research mentorship of undergraduate students, whereas others attend more to the professor's record of mentoring graduate students. Others do not count mentorship toward tenure and promotion. Thus, some professors who have an incentive to be thoroughly committed to undergraduate students may not have time to devote to research. Other universities and departments conduct evaluations that attend primarily to the quantity and quality of research publications a professor has, so such a professor might not have time to mentor undergraduate researchers. But within each type of institution and department are professors who choose to work with undergraduate students on their research for all kinds of reasons—and we'll help you find those faculty.

Undergraduate research mentorship is a great way for professors to combine their focuses of research and teaching. Some professors run research groups or labs with both graduate and undergraduate students in order to bring together different perspectives and skills, create a chain of mentorship, and help students succeed. Some professors prefer to work with undergraduate students rather than graduate students because they find that their own research projects and the field of research in general benefit from the fresh perspectives of newer researchers. Some professors themselves had impactful faculty mentors when they were undergraduate students and want to pay it forward. Still other professors are looking for undergraduates to do the technical work of research and/or to make sure that they have enough researchers to get all of the work done. Many professors—like us—choose to work with undergraduate researchers because of all of these reasons! There are also professors who choose not to work with undergraduate researchers. See Textbox 4.2 to learn more about why, as well as to learn answers to other questions you may still have about the professor life.

Now that we've shared what a professor is and does, let's look at what it takes to find a professor who can be a research mentor.

WHAT IS A MENTOR?

What can you expect from a research mentor? As with professors in general, what a research mentor does depends upon the nature of the research, the mentor's view of work with undergraduates, and the culture of the mentor's discipline, department, and/or college. Several researchers who themselves mentor undergraduate students have written about what good faculty mentors do and can do. A brief review of these mentoring strategies will help you know what you might want to look for in a mentor, and help you ask and answer questions so that your mentor can offer you the best guidance possible. Some researchers who study mentorship, such as Buffy Smith from the University of St. Thomas in Minnesota, focus on mentorship for underrepresented students. Based on her research interviews with students and mentors, Smith (2013) explains how faculty mentors use the long-term goals of their mentees to motivate students to develop their writing skills and to learn how to seek research opportunities. Your faculty mentor can best serve you if you share your goals for later in college and after college, and then ask questions about how you can use research to achieve them.

Other mentorship researchers, such as Gina Wisker at the University of Brighton in England, focus on research mentorship for student-led projects like honors theses. Based on the research literature on student mentorships

Textbox 4.2: Q & A About Profs and What They Do

Q: What are some typical reasons that people want to become professors?

A: A great question to ask professors in office hours, as the reasons vary greatly! A professor might be devoted to students, teaching, research, writing, their discipline, or a life of the mind. A professor may have been drawn to the flexible schedule, the stability of tenure, or the prestige of the position.

Q: What happens if a professor doesn't earn tenure after the probationary period?

A: First, know that there are many reasons that a professor might not earn tenure. A professor may find that their work—be it teaching, advising, research, or outreach—is not valued within their department, discipline, or college. The professor may then find a good tenure-track position elsewhere. Discrimination can also prevent some professors from earning tenure. You can learn more about discrimination in tenure and promotion in the section "The Role of Underrepresented Faculty and Intellectual Areas" later in this chapter.

Q: Why do some faculty *not* have opportunities to work with undergraduate researchers?

A: Just as with reasons that faculty *do* work with undergraduates, there are many reasons why faculty might *not* work with undergraduates, often including structural barriers: preference or need (for evaluation purposes, for example) to work independently rather than collaboratively, little money available to fund undergraduates, advising taken up by graduate student mentoring, and specific discipline, department, or college values.

Q: What is sabbatical, and how can it impact my working relationship with a professor?

A: If a professor is *on sabbatical* they have been granted a semester or more of paid leave from teaching, committees, etc., usually to work on research. Some faculty will be unreachable during a sabbatical, others will be reachable only electronically or by phone, and still others will remain on campus. Some faculty don't have sabbaticals or they have to pay for them. If you have a new research mentor or are trying to reach a professor on sabbatical, make sure to ask about how the professor's leave might impact your research relationship.

We also encourage you to ask your own professors these questions.

and her own research interviews, Wisker (2005) advises research mentors to collaborate with their research mentees to set communication strategies, expectations, agendas, and timelines. Faculty mentors may have different strategies in communicating with their students. Some prefer to meet on a regular basis with their research students (e.g., weekly) while others take a more hands-off approach and check in more intermittently. In your initial meeting, ask the professor about these expectations for communication. If one of these communication styles suits you better, your preference may factor into your decision about a research mentor. For example, do you want more of a hands-on approach from your mentor or do you prefer to work more on your own with fewer face-to-face meetings? You may also have to adapt your communication strategies if your professor's philosophy does not match up with yours. If you are unsure about how to ensure effective communication, ask your mentor so that you won't be left wondering what is expected of you or when, how, and how often to communicate. See below for sections on specific tips for communicating with faculty in meetings and via email. One way to evaluate whether a faculty mentor will be a good mentor for you is to speak with the students who currently work or have worked in the past with the professor. Ask them questions about the professor's communication style, their management techniques, and the skills that the students have been able to garner from this relationship.

A good research mentor should also help you realistically manage your time and develop your writing skills (Wisker, 2005). In order to make the most out of what your research mentor has to offer in time-management and writing skills, read Chapters 3 and 5 of this book and use them to prepare questions for your mentor about how to take your skills to the next level for research work. According to Wisker (2005), the job of a research mentor is to "[h]elp students to help themselves and each other" (p. 159). Your mentor should be offering plenty of guidance, but not necessarily telling you directly how to undertake your work. Remember that with research, there is no right answer. If your mentor does not offer this guidance naturally, you may need to ask questions to encourage more guidance. Much like this book, a good research mentor will lead you to ask and answer questions on your own. In the next section, we'll look at finding professors who will be good research mentors.

FINDING ENGAGED FACULTY

In the previous section we described what a research mentor is. In this section we describe how to find the best research mentor for you—one who is invested in your success and is willing to go that extra mile for you. An engaged faculty mentor is the difference between a good research experience and a great research experience. So how do you find an engaged faculty mentor? First, make sure the professors who teach your courses know you early on. Meanwhile, make sure that you are the best possible candidate for working with faculty on research. Before taking you on as a research student or mentee, a professor might want to know if you will meet agreed-upon deadlines, be on time for and attend all meetings, complete necessary research components (e.g., reading, data collection, etc.) correctly and in a timely manner, respond promptly to requests and questions from you advisor, and stay committed to the project. Begin thinking of ways you can let potential mentors know that you will follow through with these responsibilities of a student researcher. Potential mentors might look at your academic record and experience, using your GPA and the courses you've taken to determine your preparation and potential for the research. They also might ask you questions to learn about your ability to meet certain criteria in other areas (e.g., group work in your courses, time management with extracurricular activities), so be prepared. A potential mentor might ask other professors about your academic work, so keep that in mind throughout your time taking classes and interacting with faculty. Reading this book will definitely enhance your research potential as well. In this chapter, we'll also help you plan for communication when things go wrong—know that they may, and that's okay!

Think about your own research interests and then you can look up faculty with similar interests. Getting to know your professors through classes is a great start. You might contact a professor with whom you've taken a class that you found interesting to see if the professor offers research experiences. You can also ask your academic pre-major or major advisor to recommend faculty based on your interests. You can look at professors' LinkedIn profiles or their professional social media websites. You can look at faculty pages on your college's website where you might find descriptions of faculty research or links to faculty curricula vitae (CV). A CV is the story of a professor's academic life, structured like a long resume. Looking at a professor's CV can help you answer questions like:

- What research is the professor working on?
- What research has the professor worked on in the past?
- What was the path to the professor's current position? Where did the professor go to undergraduate and graduate school?
- What has the professor written and presented?
- What and how many student theses or independent studies has the professor supervised?
- Has the professor presented or published with students?

To get a more personal sense of what it's like to work with a particular professor, you can talk to students, undergraduate and graduate, with whom the professor has worked. Other students can give you a sense of the professor's expectations and how receiving research mentorship has helped them with college or enhanced their academic experience. Most professors have a website that provides information about their research interests, their research group, and ongoing and completed projects. Research websites can give you important insight into whether the research this professor does is interesting to you and, just as importantly, if this faculty member might be a supportive mentor for you. For example, a website that features students and specifically mentions undergraduate research as a priority is a good sign that this professor values undergraduate research.

Another great way to find a research mentor is to use the research resources on campus. Many colleges and universities have an office of undergraduate research or another office that supports student research on campus. Ask around or look under the academic or student services link on your college's website. The staff at these offices can help you identify a professor who is doing research in your area of interest. A research or reference librarian at your college library can also be a great resource. You might be able to make an appointment or stop by a research or reference desk at your campus library. A librarian can help you find the professor's dissertation as well as other published work. Make sure to follow the steps above once the staff member has identified a potential professor for you—check out the professor's CV, talk to their students, see what classes they teach—before you send an email.

If you cannot find a faculty mentor at your school because the professors you have contacted are not working with students or because there is no one who is doing research on the topic in which you are interested, a good option is to try to find a mentor at another school. There are faculty at universities around the country who may be working on a topic of interest to you and would be happy to work with an undergraduate scholar. Find a professor in your department to get some direction about whom to contact. Look at the websites of professors at colleges near you to see if anyone is doing work that excites you and send an email inquiring about research opportunities. See the Email section below for tips about how to communicate with faculty about research opportunities. In addition, there are many really great opportunities for paid summer research experiences at other institutions. See Chapters 2 and 3 for more information.

Email

With some potential research mentors, email may be one of your first means of contact. Even if the professor wants to meet you in person initially, you should use email as a way to set up this meeting. We recommend that your email be no longer than one short paragraph—your professors are busy and may get hundreds of emails a day. You can start by introducing yourself briefly—name, year in school, major or intended major—and stating your interest in research. Then, ask about the professor's schedule or office hours. Show your commitment to working on research by accommodating the time that the professor has already set aside to work with students. Once you have confirmed a time to come to office hours or another time for a meeting, you might also ask if the professor would like to be reminded on the day-of. Some prefer a reminder (especially if you've made the appointment well in advance) and some don't. When emailing professors, pay special attention to wording and format. Find tips in Textbox 4.3.

Office Hours and Appointments

Office hours are a time when you can walk in and talk to your professors about any ideas or questions you may have. Office hours are a great resource if you have questions about class that you can't find answers to on your own. You can also build relationships with professors through office hours. Many students are intimidated by the one-on-one nature of office hours. Remember that your professors were once in your position! It will also help you if you come prepared with an idea of what you might expect, so this section will help you prepare. Chapter 6 can also help you find professors who make you feel welcome.

Some professors may not have set office hours. If you have taken or are currently taking a class with such a professor, check your syllabus for guidelines on how to set up an appointment. You might also find this information on the professor's website. If you're still not sure how to set up an appointment, either email the professor or see them before or after class to verify whether they hold set office hours and to ask how to set up an appointment. Department administrators are also good resources for finding out when professors are generally available; you should be able to look up the administrator's contact information on the department's website.

Upon arriving at the professor's office, be sure to introduce yourself and remind the professor what you are there for (e.g., "I want to find out about research opportunities"). As you get started with research, a good question to ask professors is: "What current or upcoming research projects are you working on?" If you've found a professor who works with undergraduate students on research, it is likely that they will have some research tasks available for you. If not, you can find out what you might do to get involved later on, or ask about other faculty with similar research. You can also ask how to get on track for more independent projects like a thesis, capstone, or independent study (see Chapter 5 for more information on these options).

Textbox 4.3: Tips for Email Wording and Format

The subject line: Be specific! Your professors might receive hundreds of emails daily, so "Meeting," "Research," or "Office Hours" will not stand out or help them easily find your email later. You can be specific by including your name and your status as a student followed by the topic and date(s) if applicable. For example: "Hannah Franz, Undergraduate Student Interested in Research Opportunities" or "Do you have available office hours the week of 9/29 to meet re: undergraduate research opportunities?"

Addressing professors: We recommend that you keep it formal unless explicitly told not to: "Dear Dr. . . ." or "Dear Prof. . . ." In person, you can ask professors whether they prefer the title Doctor or Professor. These conventions depend on the culture of the college and of the discipline, as well as on individual preference. In addition, some professors may not hold a doctorate. Some professors are more informal and will want you to call them by their first name. When in doubt, however, use "Dr." or "Prof."

Length: For the first email, keep it brief! Pollak (2012) recommends five sentences as a rule of thumb. Later on, more lengthy emails may or may not be acceptable. Once you begin working with a professor, ask how often and for what purposes they like to be contacted by email and whether or not they prefer to have non-logistical conversations over email.

Closing the email: Remember to thank professors for their time. We like "Sincerely" for closing, but many academics use "Best."

Here's a sample email using these tools:

Subject: Student in Statistics class inquiring about office hours
Dear Prof. XXXX,

I am a student in your Statistics class and would like to meet with you to go over the last exam. Do you have available office hours this week so I can come by and ask questions about the incorrect items on my test? Thank you for your time.

Sincerely,
Your Name

TEXTBOX 4.4:
SAMPLE QUESTIONS YOU CAN ASK IN OFFICE HOURS AND APPOINTMENTS

Get to know the professor:

- What courses will you be teaching in the future?
- Do you work with undergraduates on research?
- What research projects are you working on or planning for?
- Why did you want to become a professor?
- What made you interested in your research topic?
- Why did you decide to work at this college?
- Give them a compliment on their class or their work! (e.g., "I really enjoyed the last paper you published" or "That class when you talked about 'X' really captured my interest")

Learn about their research with students:

- Are there classes I need to take or other prerequisites for working with you on research?
- Are there other students who work with you that I should meet with?
- Are there readings that you recommend I complete before I get started?
- How many hours a week would I need to commit? Would research hours be at a particular time or place?
- Do you have regular research meetings with students? When are they?
- What work do you require research students to turn in to you and how often?
- What are you looking for in a research student?
- Do undergraduate students generally initiate the topics of their research or do they usually work on a project of your choice?

Remember to show your enthusiasm! Most professors love the research topic on which they work, so they will respond well to someone who shares their passion for the area. Once you learn about possible research opportunities, you can ask questions about expectations and preparation for research. For examples of these and other types of questions, see Textbox 4.4.

Answers to the sample questions will help you get to know a professor and their work, including their research. If you are already considering working with the professor on research, the second set of questions will help you decide whether you have the time to commit to a particular research

Activity 4.1: Prepare for Your First Research Meeting with a Professor

1. Research a professor that you've had in class or that you may be interested in working with. Find the professor's CV and check out their website.
2. On their CV, identify 2–3 works or projects that the professor has written or created. Look for recent works and works that tie to your interests.
3. Find the works through your college library. Read them and see how they fit with your interests and goals.
4. Email the professor and plan to visit them at office hours.
5. Prepare your own CV, resume, or a statement of your skills and experience to bring to the office-hours visit.
6. In office hours, share your goals and interests in research, keeping in mind what you've learned about the professor's work.
7. Ask about current and upcoming research projects you might be able to work on. Be ready to be flexible in your interests based on current projects and available opportunities.

project. You might not get to all of these questions at one meeting, so use a potential research mentor's office hours regularly. Office hours and other research meetings will help you sustain a relationship with a professor, making for extra opportunities to learn about research and potentially to receive a strong recommendation letter!

When you've learned about the time commitment for a particular research project, use your calendar from Chapter 3 to determine how you might fit in the research while staying successful in your classes and maintaining good health. Be honest with yourself and your professor as to how many hours per week you are able to fit in to do the necessary work. We have seen students who overcommit to classes and extracurricular activities, and their research tends to be the work that falls behind. Make sure ahead of time that you will be able to succeed in all aspects of your college life. Then, use your calendar to schedule research time, including lab meetings and office hours, as fixed commitments.

Activity 4.1 can help you prepare for an initial meeting to get started with research.

Scholar Sara Taylor describes the process of communicating with her faculty research mentor after hearing about the mentor's research group through a friend and visiting the mentor's research website:

After I found a lab that was interesting to me, I waited until about halfway through the first semester of my freshman year to contact her so I had time to get used to my classes and college life in general. To get a meeting with one of the professors running the lab, all it took was an email. I was so excited and nervous going into the meeting, ready to talk about my research in high school and make my case for a spot in the lab. Instead, the professor (who has been my research mentor for three years now) explained what kind of research the lab conducted in order to make sure it fit what I was interested in. I have now been working in the lab for 3 years and am currently conducting an honors project. It still amazes me that it all started with an email and a 20-minute meeting.

WORKING WITH FACULTY

Finding a Research Topic

Research mentors who regularly conduct research in a given area can provide the necessary expertise to help you learn about the topic and can oversee your research project. Many professors who work with undergraduates will have research projects that you might be able to work on. Some professors might also supervise a more independent project that you take the lead on. Independent projects can take the form of an honors thesis, capstone, independent study, or summer research experience.

In all cases, you will need to remain flexible on the topic based on factors such as faculty research responsibilities, ethical guidelines, and practical considerations such as what budgets will allow for. Flexibility is part of research—these factors impact both professors and students! At smaller colleges, flexibility may be even more important as you find research mentors, because you are less likely to find faculty whose research aligns with your specific interests. Such flexibility will be important even if you are working on a more independent project, because professors do not always supervise projects that are outside of their area of expertise. For professors seeking tenure and promotion or simply planning their time wisely, it makes the most sense to supervise projects that align with their own current research projects. In some cases, you may need to meet your research mentor in the middle to find a project topic that satisfies you both.

If you can make an argument that your research is important and interesting, and show that you have a strong grasp of the previous work on the topic, the professor will be more likely to support a project you really want to do. It will help if you can justify your interest by knowing how your research is unique and what kinds of problems can be solved as a result. You will be committing a lot of time to your research project, and it will be much more enjoyable and you will be much more successful if you are excited about the topic. (The three of us all study topics about which we are deeply passionate, and this dedication and enjoyment is key for our surviving the long hours that successful research often takes.) It is important to settle on a topic and a research plan with your research advisor before you begin the time period allotted to your project. If you take the lead on choosing a project topic in which your advisor does not have a lot of expertise, this can be a positive experience because you and your advisor are learning together and discovering a new field for both of you, but it can also take longer to get to the point where you both fully understand the previous research and can design an appropriate project.

Even if you cannot do the exact project that you would like to, remember that you will still be gaining valuable mentorship and learning useful research skills, so keep up your enthusiasm for the work! Just like most professors, you probably have many interests and questions that could spark research. You will work on only one of them now, but once you gain research experience, you will have more opportunities to explore further issues and questions.

Sharing Authorship

Working on research with a professor means you may have the opportunity to present your work at a conference or to submit it for publication in a journal or a book. These opportunities are great for learning about the processes entailed in presenting research, which are often nebulous but can have a number of rewards. One reward is simply the ability to share what you found in your research with other scholars or community members. You can also potentially make a lasting contribution to your field. Presenting research in these forms can yield feedback to help you refine your research project, help you make contacts with important people in your field, and increase your competitiveness for graduate school. So take advantage of presentation and publication opportunities. When presenting and publishing from research you have worked on, you will want to get credit for your contributions, as will the other researchers involved. See Chapter 5 for more about presenting your work and writing for publication.

When working with a professor on a research project, ask early on about the possibilities for presentation or publication. This topic might arise naturally when you and your advisor agree on the type of research you'll be conducting and the scope of your project. If your advisor doesn't bring up presentation or publication, ask about the possibility. For every presentation or publication, authors are identified as having contributed to the work; sometimes there can be two or three authors (as in the case of this book), but we have seen papers that have more than 20 authors listed! When presenting at a conference, publishing a paper or book, or registering a copyright or patent, researchers need to decide who will be listed as an author on the work. Factors for determining authorship differ by discipline, and sometimes by publication outlet, even within the same field. Although most fields provide official authorship guidelines, authorship decisions are sometimes subjective, dependent upon individual views on ethical authorship. Early in the research project, get a sense of what authorship means to your mentor or advisor and which guidelines, if any, they adhere to.

In addition to deciding who is an author, researchers need to decide in what order to list those authors in the publication or presentation. The order of authors communicates the researchers' relative contributions to the work. The term *lead author* may take on different meanings; the lead author may have initiated the research question or idea, may have conducted most of the research, may have done the bulk of the writing, or may have provided the infrastructural support (e.g., lab space, money) necessary for the project. Authorship order means different things in different disciplines, so ask about guidelines relevant to your discipline, project, and advisor. In some fields (e.g., psychology), the lead author is listed first, whereas in other fields (e.g., physics), the lead author is listed last, and still in other fields (e.g., computer science), authors are often listed alphabetically.

It is important to make sure that your contribution to the research project will be honored in an appropriate way so that researchers in your field and potential graduate schools can identify your contribution. In that initial conversation with your advisor, you should ask about how decisions are going to be made about who will be listed as an author and in what order, as well as who will be responsible for presenting the work or who will be the main contact for the journal, book editor, or publisher. Remember that research is dynamic and changing, so continue to ask questions about presenting, publishing, and authorship throughout the research project.

If you are working on an honors thesis or another more independent project, ask your advisor for guidance on presenting or publishing your work. You will want to ask about authorship in this case as well. Often the student working on the thesis is the lead, but there are exceptions, so ask to avoid surprises from your advisor.

It is important to have ongoing and open conversations about authorship so that you are well positioned to avoid surprises that you might find unfair or unethical. Two such surprises that can occur with research publications are ghost authorship and guest authorship. A ghost author is someone who contributed enough, perhaps even a substantial amount, to a paper to qualify for authorship, but is *not* listed as an author on the paper. A guest author is the reverse: an author who is listed on the paper, but who did not contribute enough to qualify for authorship according to authorship guidelines. While ghost and guest authors may sound like extreme examples of authorship issues, both situations occur at a higher rate than you might expect. Faculty and student research team John P. Walsh and Sahra Jabbehdari surveyed 2,300 lead authors on published papers and found that 33% of the papers had a guest author, 55% of the papers had a ghost author, and almost half of the ghost authors were graduate students (Jaschik, 2015; Walsh & Jabbehdari, 2015). As with authorship guidelines, authorship issues vary with discipline, so research the rates in your areas of research. For example, researchers Seeman and House (2010) surveyed 600 faculty in chemistry about their crediting practices and found that faculty were more likely to give credit—in the form of authorship or an acknowledgment—to their own student than to another faculty member's student, when contributions from the students were equal.

Take heart that, as an undergraduate student with awareness about common authorship issues, you are ahead of the game (and ahead of many graduate students) and can anticipate these issues before they arise. So again, ask your advisor or mentor about authorship throughout the research and writing process. For more resources, see The American Psychological Association's guide for student researchers (www.apa.org/science/leadership/students/authorship-paper.aspx).

Navigating Challenges

Working with a professor offers many rewards, but, as with any form of collaboration, misunderstandings, inability to follow through (on your part or the advisor's part), and other challenges may arise. These challenges will be less overwhelming to you if you are prepared for them to occur. One way to prepare is to talk to other students who have worked with your advisor or with other research mentors. For example, ask more advanced research students questions about what happens when they or their advisor are unable to meet deadlines, respond to emails, attend meetings on time or at all, understand expectations, or stay committed to the research project. If it seems like your advisor is not meeting their responsibilities, continue to ask them questions to clarify those responsibilities. You can ask, for example, "So that I can plan ahead, when should I expect feedback from you on my drafts?" "Should I plan to continue the 1:00 p.m. Tuesday meeting time for the rest of the semester?" "If I have a question, do you prefer that I email you or come to your office hours or something else?" If it seems that you are unable to resolve differences between you and your research mentor, sometimes it's best to cut ties and try to find another mentor. As scholar Bailey Johnson (a pseudonym) comments:

> My first encounter with a research lab came from one of my favorite professors. She asked her teaching assistant (TA) to reach out to me and give insight on what duties in the lab are. That night, I wrote an email to my professor with a list of questions about being involved in the lab, and weeks went by without a response. I even confronted her after class inquiring if she read the email. She replied, "oh yeah, I've seen it. I'm just really backed up on my emails." As time went by, I only felt more and more awkward about the situation since I was present in her biweekly class and she never mentioned the lab position. I reached out to upperclassmen and other advisors for guidance, not knowing how to confront this situation respectfully and efficiently. I finally got the courage to stay after class and just ask her what was going on. She made it seem like it was my fault and that my lack of communication was the issue. I couldn't believe it! I ended up turning down her offer, and even applying to participate in a new lab. Now, I'm proudly enrolled in a research lab with another faculty mentor where the communication is 100% better and I feel completely supported.

If it seems like *you* are not meeting your responsibilities, ask your mentor questions to make sure you understand what is expected of you. Try re-evaluating your time management and other skills needed for the project: Are there skills you need to develop to meet your research responsibilities? Let your research mentor know so that you can get guidance. Review Chapter 3 for questions and resources to help you develop time and energy management for research, and read Chapter 5 to learn more about research writing. You can also re-evaluate your enthusiasm for the project: Have you begun to prioritize other commitments? Why might that be? Why were you interested in the project in the beginning? Even if the topic or scope of your initial research interest has changed, consider the benefits of staying committed to the project—what skills, contacts, and recommendation letters are you gaining? The next section will explore ways to add multiple perspectives to your research, which may help reignite your enthusiasm.

THE ROLE OF UNDERREPRESENTED FACULTY AND INTELLECTUAL AREAS: THE RISE OF RESPONSIVE PROGRAMS

Making your research interdisciplinary is one way to ensure that your focus is unique and significant. In particular, adding a focus from an area that is intellectually underrepresented in the academy can bring important perspectives to your work and help you conduct research that is ethical, representative, and responsive to existing questions and needs. By *an area that is intellectually underrepresented in the academy*, we mean an academic area that addresses inequities in perspectives and in how those perspectives are presented. Such intellectual areas include Africana Studies, African American Studies, Asian American Studies, Ethnic Studies, Native American Studies, Latinx Studies, Sexuality Studies, and Women's and Gender Studies. These areas aim to break down the division between the researcher and the researched, and to use research to draw directly on the perspectives of the community of focus. Research in the academy is becoming increasingly interdisciplinary (Basken, 2012; Petrie, 2007). If you are not already working with a professor in an underrepresented intellectual area, see Activity 4.2 to strengthen your research potential and create a larger base of research mentors.

In any research area or discipline there are many benefits to working with professors who are from backgrounds that are underrepresented among college faculty. In Chapter 1 you read about the meaning of *underrepresented students*. Researchers Trower and Chait (2002) at the Harvard Graduate School of Education collected past and current data on faculty at U.S. colleges by rank, race, and gender. They found an even greater underrepresentation of professors than of students, for both faculty of color and female faculty. Moreover, the overrepresentation of White male faculty increases as faculty rank increases. At the beginning of this chapter, you read about faculty rank and the tenure and promotion process. Faculty from underrepresented groups can face discrimination in this process when, for example, the perspective of their research is undervalued or they are asked to serve, as a token member, on more committees than others, cutting back on the time they have available for other projects (Trower & Chait, 2002). Faculty without tenure and aspiring faculty from underrepresented groups are also less likely to receive research mentorship. Thus, if you receive mentorship from an underrepresented professor, you might have the opportunity to learn about how to navigate challenges in college and graduate school. You can also learn why they wanted to be a professor and how you can contribute to more equitable representation in and across colleges. Chapter 6 can give you more information about this process.

With the overrepresentation of White male faculty, especially in senior roles (Trower & Chait, 2002), it is important to consider what it means

ACTIVITY 4.2: WIDENING YOUR RESEARCH PERSPECTIVES AND MENTORSHIP OPPORTUNITIES

1. Find a department or program at your school that is an underrepresented intellectual area, such as Africana Studies, African American Studies, Asian American Studies, Ethnic Studies, Latinx Studies, Sexuality Studies, or Women's and Gender Studies.
2. Read the descriptions of courses offered in that department or program. What could one or more of these course topics add to your current research interests?
3. Find and review the CV of a professor in the department or program. (See Activity 4.1 for questions to help you learn from a professor's CV.)
4. Email the professor and make a plan to visit them in office hours.
5. Prepare questions for your office hours visit to learn more about the professor's work and how their work might align with your interests. See Textbox 4.4.
6. In office hours, make a plan to take a course with the professor, visit their office hours again, and, if possible, get started with them on a research project.

to mentor and be mentored across gender and cultural lines, particularly for underrepresented students. Rockquemore (2016), who studies faculty development and leadership, explains that White male mentors can and should meet the needs of mentees of color. Rockquemore provides a mentoring map that mentors and mentees can use to focus on specific needs related to academic and professional advancement (National Center for Faculty Development and Diversity, 2011). Cross-cultural mentoring brings benefits and challenges. To research student perceptions of mentoring, Dahlvig (2010) interviewed African American female students at a predominantly White institution and found that with White mentors, unlike with African American—especially female—mentors, the students felt they had to "explain everything" (p. 388). At the same time, the students built meaningful relationships with White mentors while they were isolated from their White peers (Dahlvig, 2010). You can find an entire volume of the *Australian Journal of Career Development* (Vol. 38, 2011) dedicated to reviewing research in this area and offering strategies to maximize the benefits of these relationships.

CONCLUSION

As you find and work with research mentors, you will be able to develop a variety of useful skills. The relationship between mentors and mentees aids in the process of thinking, writing, and presenting that is so crucial to academic research. The research experiences you have as an undergraduate can lead to opportunities to write and present your work. In Chapter 5 you will read about how to use your communication skills to share your findings with the world.

Writing and Presenting Research

One of the most challenging yet most enjoyable parts of the research process is presenting your work to other people. While there is benefit in learning for your own personal and intellectual growth and enrichment, learning everything there is about a new topic or conducting your own exciting research is made even more dynamic by sharing it with someone else. Why keep all that you've learned to yourself? Strong research presentations in the social science (e.g., psychology) and natural science (e.g., biology) disciplines, for example, explicitly present the reason for your research, explain the work that you have done, put this work into the context of the field, and accurately summarize your findings for appropriate audiences, conveying the importance of your work. If you are conducting research in the humanities disciplines, you are going to present the reason for your work and how it fits into the literature, but you are going to focus more on the specific material that you examined in your work (e.g., texts, interviews) and summarize your argument based on this material. For a presentation in the performing arts disciplines (e.g., music, theater), your presentation will likely consist of you reading your writing aloud, performing music or dance, or presenting visual arts such as drawings, photographs, or other media. For the arts in particular, the presentation itself is an integral part of the academic work. As you can see, although the goals of presenting your work are similar in that you are telling or showing your audience what you have spent your time doing, each general area of study has different presentation formats, and they will overlap. You should have your research mentor help you identify what seems most fitting in the way of presenting within your area and for a particular audience. You might have the most important, creative research study in the world, but if your presentation is not accessible to your particular audience, the impact and quality of your work may go unnoticed. This chapter will discuss the importance of presenting research as part of the research process.

In this chapter we'll emphasize comprehensive writing when completing a literature review and reporting research findings both in classes and in research experiences. Writing is an art and a skill, and most academics will agree that it is a lifelong process that you should constantly work to improve. We know we still are.

THE NOTION OF AUDIENCE

Whether you are writing a literature review or research paper for a college course or presenting the results of your own original work at an academic conference, a good place to start the writing process is to think about your audience. Who will be reading, watching, or listening to your work? Who will be evaluating it? What is the outcome that you would like—an A for a class or publication in a scientific journal? Who else in the world do you want to reach? Your classmates? Your greater community? A high school student who would benefit from your insights as they prepare to apply to college? Someone who hasn't even been born yet? These questions will help you begin the writing process before you even start that first sentence. It can be a challenge to write to different audiences yet preserve your own voice. Differences in culture, race, sexual orientation, and background may add to that challenge, as the culture of your reader may be very different from your own. Questions about such challenges can help scholars, particularly underrepresented scholars, decide what research communities they want to be a part of and discern whether they will be more readily accepted into the community or will have to work to navigate toward acceptance—and if that navigation is worth it.

Nikki Giovanni (2007), a world-renowned writer and professor of English at Virginia Tech, writes about this notion of audience and demonstrates that writing can be audience-centered in her poem entitled "A Higher Level of Poetry." Giovanni states: "There is really only one thing to say to young writers: Know who you are writing for and to." She is explicit in the poem that her intended audience is the women of her family and others who maintained communities through church and civic participation during the civil rights movement. Giovanni explains, "knowing who you want to be proud of you can make all the difference in the world." She isn't against others reading her work, but she is very clear that "I want my Grandmother and her friends to look back at my work and be pleased" as well as all of the Black women who had stood up to state-sanctioned oppression before her and knew that "if they didn't stand then all that death was in vain." In this way, Giovanni assures that you will know you have done all you can do and, "you can be proud of your work" (p. 103).

Scholar Nathanael Paige exemplifies Giovanni's message in his method to get writing started:

> While writing my paper it also helps to imagine that I am presenting my paper to a population of people that will benefit the most from my findings. This particular strategy allows my paper to become interesting to read, as I am reminded that the content of the paper can positively impact someone.

Here are some questions so that you can do the same:

- Who do you want to be proud of your work?
- What makes you proud of your own work?
- How will you know when you have done all you can do?

Such questions are particularly important to help you with the adjustment period that many scholars experience transitioning between high school and college writing. Many students report that they are hesitant to begin writing papers until they discover that you really just have to get started—there's no one right way to do it, and expectations will vary from class to class and professor to professor. As scholar Ebi Doubeni explains:

> The biggest difference in high school writing to college writing is the style of writing. In high school when a paper was assigned the teacher would go over the basic 5-paragraph formula (Introduction, 3 body paragraphs, and a conclusion) that the students were supposed to follow. When I first came to college I learned quickly that it was no longer an acceptable form of writing. Each professor had their own personal preference for writing and wanted a more complex form of writing that did not fit into the 5-paragraph formula. During my first semester and even more so now I learned that going to office hours was crucial in helping me to write papers. When I started to go to office hours, I would ask my professors what were their expectations in their students' writing and what type of writing would get an A in the class.

An important point to keep in mind as you write for research is that writing for a college audience and for a research audience is much different than writing for a high school audience. Although we often share this information with our students, they are often surprised at how big this difference is. The good thing is that, in reading this section, you will have a head start in knowing these differences and can master writing early on. The first thing to note is that being a good writer in high school does not necessarily translate to being a good writer in college and beyond, although understanding what college writing necessitates and working hard on this new style can help you be a successful undergraduate writer. In interviews with high school English teachers, Ms. Franz found that high school students might earn A's on writing assignments when they follow general formulas for argumentative or analytical writing (Franz, 2016). Meanwhile, in her interviews with college professors who teach writing courses, she found that college students might earn A's on writing assignments when they create their own original claims and follow the writing guidelines of a particular disciplinary area, such as the sciences, the social sciences, or the humanities. In this chapter, we discuss the genre of research writing generally, but also refer to disciplinary differences in this genre. One professor may mean one thing by "research paper," another professor may mean another, and your high school teacher may have meant something different altogether when assigning a research paper. You can always ask your professor or research advisor if you're unsure what your research paper or presentation should look like.

In high school, you don't necessarily have to think about the teacher for whom you will be writing during the writing process. Your writing is probably graded according to a rubric or guidelines based on school, state, national, or international standards and exams, such as those for AP, IB, and Common Core. Often you answer questions that are posed in a prompt and write in a grammar and style that adheres to these standards. You most likely learned a particular writing grammar and style in English class, although you may also have had teachers in other subjects who taught you writing guidelines specific to their class or subject. Perhaps following external standards, your teachers may have taught grammar and style as static—that is, that there is one way to write well in school. In college, you will learn that there are many academic grammars, styles, and types of good writing.

In college classes and in research, it is important to consider the role of the individual professor or professors for whom you are writing. Unless you are an English major, most of the time you will be writing for professors in disciplines other than the ones you wrote most for in high school. Every professor has their own guidelines for grading written work, and many may not include a written-out rubric or a specific writing prompt with their assignments. The style on which you are evaluated may differ from instructor to instructor and from discipline to discipline. If you're not sure how a professor will be evaluating your written work, visit office hours and ask! You can ask about specific writing styles, which we will discuss later in this chapter. You might explain briefly how your writing was evaluated in high school and let them know that you are trying to prepare for different types of writing in college. From a discussion of how a professor evaluates writing, you can learn about the writing and research values of your individual professor and also of the genre(s) or discipline(s) you are writing in. You might also ask to see an example of a successful paper for that particular class so you can model your writing for that assignment after a paper that received an A in the course.

CITATION AND WRITING STYLE

One thing that may differ between disciplines is the citation style that you utilize in your papers. As you know, you need to give credit to relevant sources when you are summarizing or quoting other people's work or ideas. Citation styles are series of rules that are specified by a discipline or professor that tell you how to acknowledge the sources that you present in your paper (e.g., books, journal articles, works of art, performances). Stylistic consistency allows everyone to easily navigate the papers they read within a discipline and gives credit where credit is due. Selecting a style will help you figure out the way that you use abbreviations, quotations, footnotes, endnotes, and scientific notation. For research papers in college courses and with college research projects, there are several different citation styles, or ways to give credit to your sources.

Before you begin your paper, it is important for you to find out from your professor or advisor which style is acceptable for your research papers. Ask, for example, whether your professor prefers one of the styles discussed below— Modern Language Association Style (often referred to as MLA), American Psychological Association Style (often referred to as APA), or Chicago Manual of Style—or if there is a particular style manual that they would recommend. In some disciplines (e.g., psychology), it is standard practice to always use the same style, while in other disciplines (e.g., biology), there may be more flexibility in the style in which you write. Regardless of the style, it is important that you learn the ins and outs of the style and that you are consistent throughout so that your paper can be evaluated favorably. Luckily, there are several good resources that you can use that we will describe below.

There are three writing citation styles that are most widely used. You are likely to come across one or all three during your time in college. First, the Modern Language Association (MLA) style is typically used in English studies such as language and literature, foreign language studies, and the arts. American Psychological Association (APA; used in this book) style is appropriate for social sciences such as psychology, linguistics, sociology, economics, and criminology. It is also commonly used in education, business, and nursing. The Chicago Manual of Style is often used in the humanities, fine arts, books, magazines, newspapers, and literary publications. These are general guidelines—it is important to ask your professor or advisor which style is appropriate for each assignment. There is not enough space in this chapter to go over the nuts and bolts of each style. In fact, there are whole books dedicated to each one. Purdue Online Writing Lab (OWL) provides a summary of how the styles are similar and different, available online at owl.english.purdue.edu/.

Although we each tend to have our own writing style, as you begin writing research papers you begin to make stylistic choices to best convey the information you wish to present. Style choices can include the length of sentences or paragraphs, the use of punctuation, the way the paper is organized, and the use of jargon—technical vocabulary that may not be easily understood by a reader unfamiliar with your discipline or topic—to name just a few. Submitting the same style of paper in two different classes may yield very different grades. This point gets back to our previous advice—know your audience. One great trick to help you decide on a writing style is to read what your professor or advisor has written. Find a research article or a book that your professor wrote and try to emulate that style to the best of your ability. Ask questions about where you can use your own style or add variation. That is, if your professor tends to use simple terms, avoid jargon. If your professor writes in the first person, make sure that you do the same. This way you can make sure that you not only write in the standard that is appropriate for the field but that your writing fits with the professor's individual writing expectations. In addition, it's helpful to familiarize yourself with the writing styles used in the readings your professor assigns or recommends for you. The more you read work for your audience, the better you will be able to write for your audience.

Regarding research papers in which you are reporting your own research findings, it is important to write in the style that is the standard in your discipline. To find what the required style is, we recommend looking at the webpage of that discipline's main organization (e.g., Linguistics Society of America, American Education Research Association, American Psychological Association). Of course, you should always check with your professor or research advisor to ensure that you are using the correct stylistic guidelines.

FINDING AND USING PREVIOUS RESEARCH FOR YOUR PAPER

After you have put in your due diligence identifying your audience and determining a citation style, finding previous research in the particular discipline is a good next step. Ask your professor about the best way to do so—what resource do they recommend for finding this previous work? Use this resource to browse abstracts of relevant articles, books, or other materials in your field. Reading summaries of previous work can help you see what has been done in the field and help you narrow down your topic (see information below about how to find material in a field). Once you begin discovering what has been done on your topic, you might get new research ideas and become excited about something new or shift the direction of your paper. Reading the literature on a given topic can be inspiring and can give you lots of new ideas.

Within the scope of the class, try to find a topic that fits the criterion for the paper and choose a topic about which you are curious. It is extremely important that you start this process very early so that you can run the topic by your professor and get direct advice about the assignment and your paper in particular—this will give you a definite advantage. You can ask if your topic idea is relevant to the assignment, the course, and the larger scholarly field that your topic and the course are situated in. When asking about scope, you might focus on questions like: Is the scope of my topic too narrow or too broad for the intent of the assignment? Is the scope too narrow or too broad to be feasible? If the scope is too narrow, you may not be able to find enough resources or have enough material to fulfill the assignment requirements. If the scope is too broad, it may not be feasible for you to complete the writing project in the time or page limit available. Asking your instructor about the paper topic several weeks before it is due will allow you plenty of time to follow your instructor's advice and will make you look good to your prof! Another reason to start early is that you may need some flexibility in your topic. That is, once you start discovering what is already out there on your topic, the scope or thesis of your research paper may change. If you think that making a change can improve your paper, be sure to check in with your professor first. Most professors really enjoy having students come to their office hours because they want you to achieve the best grade possible and they like talking about the subject matter. In addition, taking the time to give you advice about the direction of your paper up front may save *them* time later in the semester.

Literature Reviews

Once you've found a topic (and have had it approved by your instructor or research advisor), if you are working in the social or natural sciences, you will likely be asked by a professor or advisor to write a literature review. A literature review may be a work in itself that stands alone and can be published as such. You may have already written a literature review for your classes in high school in which you summarized an area of work on a particular topic. A literature review can also be the introductory section of a longer research paper or book chapter. If it is included in a larger paper or chapter, it is usually presented at the beginning of the work. It summarizes the relevant research that has been conducted on your topic of interest, shows that you understand the topic of interest, and provides the framework for your work. Of course, if you have picked your own topic, you will already have consulted prior research that is related to your paper. In this case, you will be responsible for reporting these previous findings in a specific format. Note that we said that you will be summarizing the *relevant* research that is applicable to your topic. One of the hardest things about literature reviews is how to define "relevant." Choosing research that is completely off topic should obviously not be done, but within a certain area it is often difficult to choose what research to include in your summary, particularly when there is a large body of research in a given area.

You might find that there is no relevant research on your topic. Innovative researchers who are pioneers in their field often have to pave a way to research new areas. If you are writing on a topic that is pertinent to underrepresented scholars in particular, you might have difficulty finding previous research on your topic in a field dominated by scholars from other, more represented backgrounds. When you have this difficulty, write comparatively. That is, describe what has been found in other areas and make hypotheses about what you think you should investigate in your area based on the research that you have found about, for example, culture, racism, social class, and discrimination in another area.

The organizational format of a literature review varies based on the discipline, but generally will involve a summary of the research and a critical (meaning you evaluate the sources) synthesis of that work. Your literature review should be organized in the form of a critical discussion in which you give a new interpretation of previous work, trace the progress of research in a particular field, or talk about debates or controversies in the field, just to name a few approaches. Most literature reviews end with a discussion about gaps in the literature or a research question of interest. Although the organization of a literature review varies to some extent by discipline, most reviews are organized by idea or thesis, not by the sources themselves—that's what an annotated bibliography is. Looking at examples of literature reviews in your particular field can be extremely helpful.

The synthesis and organization of literature reviews take a while to get used to. If you did most of your writing in high school in English classes, you may be good at writing focused syntheses of one or two texts, but less accustomed to synthesizing a larger number and a variety of sources. As a new undergraduate student, you may also be unaccustomed to *reading* papers written in the format of a literature review. Therefore, one of the best things you can do prior to organizing your literature review is to read other literature reviews in your discipline. These can provide important context so you can see what the typical structure of a review involves and how in-depth it is. Ask your professor for good examples of published literature reviews and examine the style and scope of these papers for guidance on how to frame your work. You can also ask your professor or advisor to look at an outline of your plan for writing your literature review. Ask the professor about how much should go into each section—which parts should you focus more on? Getting these answers up front can save you a lot of time and energy later on.

Using The Library

In addition to using your research advisor or instructor as a resource to help you construct a high-quality literature review, another great resource is your library. With so much accessible online content these days, it may seem a little old-school to walk into your library and search through the stacks for the books or articles you need. There are many different ways, however, in which you can use the library. A good place to start is to find a librarian who specializes in research. At your university, you may find that there are specific librarians who are experts in certain disciplines, so you want to seek out the person who can be the best resource for your particular topic. Librarians are great resources because they are trained to know which resource is best for you given your individual needs. They also have institutional experience, such that they may have worked with students who have taken your class before or worked with your professor; therefore, they may be able to give you advice that is specific to your assignment. Most librarians will require you to make an appointment—be sure to do this when you are first beginning to think about your topic, rather than once you have already started writing the paper. Research librarians are experts at looking up materials for a particular topic and can show you how to use databases that will help you find these materials. Ask them what strategy is best to find information on your particular topic and how to access the information you need. Although some work you'll be looking for is available online and in databases, there are other resources that you can access only in the library, such as government documents, maps, magazines, newspapers, and encyclopedias. Most libraries also have media resources such as films, photographs, videos, cassettes, and CDs. If you are interested in a specific item that the library does not physically have, seek out a librarian—many times the item can be borrowed for you through interlibrary loan.

Using Databases and Indexes. Research databases will provide quick access to crucial information you will need for your literature review. Although you can find a wide variety of these works by just doing a Google search, research databases provide sources that have been reviewed by experts to be reliable and current. Sources obtained from Google and other search engines may thus contain information that is unreliable, as many of these websites order materials based on fees paid by companies and on popularity rather than the authenticity of the source.

Online databases contain articles from a variety of sources, including peer-reviewed journals, newspapers, magazines, photographs and other images, primary documents, literature, and videos. Identifying the correct databases for your topic and related disciplines is an important step in choosing materials for your literature review. Most universities have a list of available databases on their websites, with information on how to use them. In most cases you can access the database by simply clicking on it. How do you choose the most helpful database? This is where librarians and your research advisors can come in handy and save you a ton of time. You can also find a list of databases by discipline. On your library's website, you will likely be able to do a search for databases that are typically used in that discipline. Once you identify the database that is right for your project, you will enter keywords, topics, authors, titles of books or journals, and dates to search for relevant works.

Grey Literature. Grey literature is often accessible through libraries and databases but not readily available through scholarly databases. Gray (2014) gives examples of grey literature such as magazines, reports, and business or society publications. Other examples include social media and oral history collections. Grey literature can give you crucial insight into communities, organizations, and current trends in society, academia, and industry.

Primary Sources. Primary sources include your own interviews, conversations, and data collected for your particular project through social media. It is important to cite these sources and to keep careful notes so that you credit work and ideas ethically and responsibly.

EDITING AND PROOFREADING YOUR WRITTEN WORK

In addition to planning ample time for your literature review, you will need to make sure you allot time (as we describe in Chapter 3) for carefully outlining, editing, and proofreading your work. Expect to write multiple drafts of any college writing assignments. At the college where we work, professors of first-year writing students have emphasized teaching their students that all writers need to take the time to revise. If you are used to writing your papers in one sitting—as on state or AP exams—you may not have had the opportunity to develop habits of careful editing and proofreading. Don't worry: If you plan for and stick to a schedule devoted to this process, you are at least halfway there! The best use of that time is to read your work out loud without rushing. If you read out loud, you will catch many more typos and other areas that need editing than if you simply read your work silently. To increase the impact of reading out loud, read to a friend who can give feedback on both the content and the style of your paper, and can be an extra set of ears to catch any mistakes. When you read to an audience, you also make the nebulous idea of "your readers" tangible. For example, writing researcher Peter Elbow has his students read their papers out loud to him as part of their revising process; Elbow (2012) explains that this process helps the writer connect to the reader and the writer to "hear the meaning come up off the page" (p. 220). When you're scheduling time for this process, know that the feedback you receive may require a substantial revision to your work, so plan the time and energy necessary to work on that final draft. Especially for longer assignments, you may have to read your paper out loud several times in order to read for different aspects or focus on different sections. For example, you may read your paper once to check that you have made your claim in a way that your reader can understand, then read your paper to check that you have included a particular component or components of the assignment, and then proofread for any typos or mistakes. Seasoned writers and scholars read over their papers many times. In addition, many campuses offer writing and editing support. We encourage you to visit the writing center and to create peer writing groups with other scholars so that you can establish a pattern of writing, talking about writing, and editing.

Also, don't be afraid to ask your professor for an additional edit. Many professors are happy to do so for students who are willing to take the initiative to ask. We want to particularly encourage you to have professors read over work that you would like to present and publish. Professors can help you take what you've started in classes and research and build on it, as research doesn't happen only outside of classes. You can use work you have already done to present and publish, as we describe in the next sections of this chapter.

PRESENTING YOUR RESEARCH IN DIFFERENT FORUMS

A crucial aspect of the research process is sharing the information you have learned with the larger research community. Doing so contributes to what is known about your topic and can help others interpret their own research findings and pose new research questions. In this section, we will talk about several forums in which you can share your research findings with others.

Conferences

Research conferences are engaging places to learn about other people's research and to share the findings of your own research, scholarship, and creative activities with peers and experts. There are all different types of conferences where you can present your work and learn about others' research. A good place to start is a local conference that is held at your school. Some institutions have a conference for undergraduate students to educate faculty and students on campus about their research. Show your friends and professors what you have been working so hard on! There are also great conferences outside your institution at which only undergraduates can present. These conferences are typically multidisciplinary and provide a supportive and fun environment for young researchers without the intimidation of bigger conferences. For example, the National Conference on Undergraduate Research sponsors several annual conferences for all disciplines and in particular is committed to research by underrepresented students. Undergraduate conferences usually have lunches and socials, which are great opportunities to meet peers from around the country. There are both national and local conferences that focus on undergraduate research, so you can likely find one that fits your needs.

In general, conferences can vary from being extremely broad to being extremely narrow in their scope. For example, if you are presenting research in the field of psychology, you can go to a general conference like that of the Association for Psychological Science, where every type of psychology is represented, or to a conference that is more specific to an area within psychology. There are conferences that represent multiple topic areas, where you would be presenting to a broader audience; these conferences sponsor several programs and presentation sessions just for undergraduate students. See our website for more information about conferences. There are a variety of regional conferences that are within a general area of the country (e.g., midwestern United States), as well as national and international conferences. There are conferences that support research in particular racial, social, gender identification, and sexual orientation areas. For example, the Association of Black Psychologists, the Association for Women in Psychology, and the National Latina/o Psychological Association are organizations that hold conferences and sponsor journals to support underrepresented scholars in psychology. The International Psychology Network for Lesbian, Gay, Bisexual, Transgender and Intersex Issues (IPsyNet) works to support the understanding of sexual orientation and gender diversity across psychology.

If you are interested in presenting your work at a conference, you should ask your advisor to help you identify the conference that is appropriate. You can also sign up for email listservs in the societies and organizations in your fields of interest, often by going to their websites and providing your email address. You will then receive information about upcoming conferences that are relevant to your interests. Presenting at a conference is a commitment of time and often of financial and academic resources, so the earlier you can start speaking with your advisor about presenting your work, the better—the professor may not even know that you would be interested in this opportunity. In addition, many conferences require that you submit a description of your presentation several months before the conference. The amount of travel time and the expenses of the conference may play a role in your decision to commit to a particular conference. Attending a conference on the West Coast if you are on the East Coast is significantly more costly in time and money than attending a local or regional conference. Particularly if a conference is held during the semester, you may not want to miss a significant amount of class time. There are many research conferences, however, and you should be able to find one that suits your needs and limitations.

To attend a conference, you may need to pay for travel, accommodations, and meals, and there is usually a conference fee. Some conferences are free or offer reduced rates for undergraduate students, particularly conferences dedicated to undergraduate research. There are many different ways to get funding for conferences. Some have travel grants you can apply for that will help offset some of the costs of attending the conference. There are often grants set aside for undergraduate students as well as students from underrepresented groups. For most of these grants you will write up a research statement and/or a personal statement (see below). Deadlines for applications for these grants are often months before the conference, so apply early.

Sometimes there are opportunities to volunteer at the conference in exchange for a reduced or completely covered conference fee. Your college may offer travel grants, so check out which ones you might be eligible for. Occasionally multiple offices or departments can give you funding. Any grants or monetary awards you receive are great to put on your resume or CV, so apply for as many as you can! Our students will many times receive funding from multiple sources that they can piece together to pay for travel and conference fees. You can sometimes obtain funding from places on campus that do not advertise travel funding, like the chair of your academic department or the head of the university research center. Your research advisor may have funding available for conference travel and may be able to help you finance your trip, especially if you have been awarded money from other sources and just need to make up the difference in costs. Take advantage of all of these sources of funding so you can travel to fun places and get research experience for almost nothing out of your pocket! Activity 5.1 is designed to help you find conferences and to prepare for them.

Activity 5.1: Finding Conferences and Academic Organizations and How to Network at Them!

Finding a Conference That's Right for You

Even if you don't plan to attend a conference in the near future, you can get some practice now finding a conference of interest in your field.

- Pick a discipline that you are interested in (e.g., English, Physics, Dance, History, etc.).
- Enter that into your search engine as well as the terms "national conference" and the current or next year. You should pull up a handful of conferences that are in your field. Look through them to see if anything is interesting.
- Search the website of the governing body in your discipline; you should find links to conferences in that area. Type in your discipline with the words "academic" and "society" and you should find a number of academic organizations.

Networking at Conferences

Networking can be daunting for undergraduates, particularly if you are going to a conference attended primarily by graduate students and faculty. Here are some tips to maximize your time at the conference without being overwhelmed:

- Do your research ahead of time: Find the names of conference attendees you might want to meet—the presenters will be listed on the conference website, and you can often join the LinkedIn or Facebook event to see who else will be there.
- Make a point to sit next to people you don't know: Get to a talk early and sit down next to someone you don't know—ask if the seat is taken and introduce yourself.
- Use research as an icebreaker: Ask people questions after their conference talks or as they are presenting their posters.
- Use nametags: At most conferences, nametags identify attendees but also where they go to school or work. Ask them a question about their university or city (e.g., how do you like living in Chicago?) as an icebreaker and proceed from there.
- Ask questions: People love talking about themselves and their research, so ask! What topics do you research? What current projects are you working on? Are you giving a presentation at this conference?

- Prepare your elevator speeches: You should have a 30-second talk and a 2-minute talk about your research. When someone asks about your project, give them the 30-second one, and if they seem interested, you can expand to the 2-minute version.

There are steps you will need to take in order to present your research at a conference. Most conferences require the submission of an abstract, which reports a summary of your research. Importantly, you do not need to report on all aspects of your work; it is probably not feasible to talk about every single part of your work given the abstract word limit, which is often 150–500 words. You can just present the parts of your research that are especially exciting or interesting or that are most appropriate for the audience at this particular conference. In your abstract you should outline the importance of your research, how you conducted your work, and talk about why your conclusions are interesting and important. The title of your abstract is important and should provide a clear summary of your work, as researchers will often choose whose research presentations to attend by the titles. There is usually an online submission process, often several weeks or months before the conference. Make sure to pay attention to the individual specifications of the conference, as they differ for each one. A selection committee made up of peers and/or experts in your field (depending on the conference) will be reading your abstract and deciding whether it will be accepted at the conference. They will be judging your abstract on the quality of your work and how interesting or important they view it to be or how it fits into the conference's theme for the year.

For most conferences, there are different ways you can present your work. You can give a talk about your research (i.e., a paper), you can have a research poster, or you can participate in a roundtable discussion. Presenting a paper involves giving a short talk about your work. You will stand in front of people at the conference and summarize your research. In most cases, there will be two to four other speakers who will be giving presentations in the same session. There will be a time frame allotted for your presentation (often between 10 and 30 minutes, followed by questions from the audience). The way we like to think about a research presentation is that you are telling the audience a story, taking them on a trip through your work. Your story should be engaging and interesting, and you must first convince your audience that it is worth learning about. A good place to start your talk is to state the main points of your argument or your research question clearly before you get into the bulk of your work. You should talk only about the parts of your work that you think are the most relevant to your audience. Do not attempt to tell them everything you learned during your research project. Keep it simple and share your most important ideas and findings only. Reiterating your points or repeating the take-home message of your work is a good idea in an oral presentation, as taking in spoken information is somewhat difficult. Make sure you know how much time you will have for your talk and practice giving your talk using a timer—going over your allotted time is extremely unprofessional and rude to the other speakers and the audience. The format of a talk differs by discipline and research area. In general, you will want to present your research question and why it is important, describe how you conducted your work, present your findings and why they are important, and offer a succinct conclusion. You should aim to show your interest in the work while maintaining a level of professionalism by rehearsing and making sure your talk is understood by others—we tend to speak more quickly when we are nervous, so practice slowing down.

At natural science and social science conferences, you will often give a talk with the help of PowerPoint slides where you can show material that will add to your presentation. If this is the case, arrive early so you can test the presentation on the computer and projector to make sure everything is working. You should always be prepared in case there are technology problems at the conference. Print out your script and slides and take backups of your slides on a flashdrive. We have been at conference presentations where the speakers have had to give their talks on the fly, without the help of their PowerPoint slides—speakers who make excellent recoveries and can easily convey the information to the audience without the help of technology impress the audience every time. When using PowerPoint, make sure you do not present too much text on the slides—you want your audience to listen to you, not just read your points on the screen. Using charts and tables in PowerPoint is a great way to overview your findings; make sure to fully explain what the figures indicate so that the audience is not left interpreting them on their own. For some humanities presentations, you will be reading a document to your audience and then answering questions after you are finished. If you are reading a paper, print it out in a large enough type size, double-spaced, so you can read it comfortably. If you choose to read it on a laptop or tablet, print out a backup copy just in case. Make sure that your reading is not monotone, but rather conveys the tone of your paper and is expressive.

Be aware that every research area has different standards when it comes to giving conference presentations. It is very important for you to check with your advisor as well as the conference guidelines to make sure you know if you should be using PowerPoint, reading your paper, or using some other format. For all areas, try to anticipate questions from your audience. A good way to do so, besides asking your advisor, is to go to other sessions at the conference and see what types of questions people ask. You can find additional tips about giving effective research talks on our website.

In addition to reading about conference presentations, you can learn different models of speaking for different audiences in different genres. Many conferences have videos available on their websites where you can watch speakers from previous conferences to get a sense of their presentation style. Checking YouTube is a good way to search for past presentations in your field. You can also learn from great orators and negotiators from Dr. Martin Luther King Jr. to President Obama. Find video clips online to get inspired.

We've focused so far on the kinds of talks that are the standard in many fields, but these formats are discipline-specific. In the visual arts and performing arts, researchers often present in the form of a performance or other artistic medium. Performing arts presentations can include creative activities in music, dance, and theater, while visual arts presentations often include artwork, photography, mixed media, and film. For performances, it is important to find out ahead of time what resources and spaces will be available to you. Will audiovisual equipment be supplied? How big will your performing space be? For visual arts presentations, you'll need to find out particulars about your presentation area. How will you transport your work and store it at the conference? How much time will you have to set up? Ask questions of your advisor and other students who have attended similar shows or conferences, as well as the organizers of the show or conference.

A research poster is another good way to present your work and get the experience of receiving feedback. In a poster session, you will be standing in front of a poster as others walk through rows of other posters with other presenters. Most conference attendees will walk around and look at the titles of the posters, stopping when they see something that catches their eye. As you stand with your poster, people will come up to you to learn about your research and ask you questions. Your poster should report on the findings of your work and should communicate information similar to what you would discuss in a paper presentation (see above). Posters will range in size (they are typically around 3 x 4 feet)—check the conference guidelines to see what the format of your poster should be.

A good way to draw your audience in is to have promotional material like handouts or goodies like candy or other swag. When they come to your poster, sometimes people will just read what you have on the poster and ask you questions when they are finished reading it. Others will ask you to give them a verbal overview of the work you have done. You should be prepared to give a carefully prepared summary that should run about 2–3 minutes. Being able to successfully present your research in a condensed way like this is a very important skill. This is often referred to as an elevator speech because you explain your work in the time it takes for an elevator to travel from one floor to another. Discussions at posters can lead you to see your research in a new way, generate new research ideas, and foster connections between you and others in your field. Talking about your work at a poster session is a great way to network, and can be less awkward than approaching strangers and starting a conversation.

Scholar John Nguyen explains:

> The project that would dominate the rest of my undergraduate research career perfectly merged my personal interests with my academic ones. The project made use of EEG, a technique that is designed to measure electrical activity in the brain. As a neuroscience student, it was an incredible opportunity to work with an instrument that people in my actual field used. My favorite part of my entire research trajectory was toward the end. Not because it was almost over, but because it was a perfect encapsulation of the highs of research—this was important work with meaningful implications. I went with my advisor to a conference in San Diego, where thousands of researchers in the same field would congregate to share their findings with colleagues. It was amazing! I got to make and print out my own conference poster, just like the ones I used to see. I got to hang it up on the poster wall and stand among all the other researchers and discuss my work. My advisor told me it looked like I had caught the research bug. I still remember her saying it, because I felt at that moment she was right. I felt this sense of energy and excitement at being able to share my findings with people who were just as interested in it as I was and hearing their feedback and ideas. I came away with a much fuller understanding of what it means to be a researcher.

When creating your poster, there are several things to keep in mind. Your text should be visible from about four feet away, so make sure it is large enough. You should aim for a minimum type size of 44 points (use Times New Roman or a similar very legible font). Your poster should be organized into sections that summarize your research in a way that presents your work in the standards of your discipline. For all disciplines, highlight why your work is interesting and important. Fewer words on the poster are better, and using tables, figures, and pictures are great ways to clearly present your results. Many people choose to use PowerPoint to design their posters and then send their document to a printer at their school or to a copy store that can enlarge the slide to fit your requirements. Typical posters cost $20–$50, so remember to include that in your conference budget so you are not paying out of pocket. Your advisor or a student in your lab will most likely have a template of a poster or a file with a previous poster they've presented, so ask for this template before starting from scratch. It's a good idea to print out a few dozen copies of your poster sized down on regular paper with your contact information so you can hand these out to people who stop by to speak with you. Giving your audience something they can take away is a good way to network and get your name out there.

Scholar Sara Taylor describes her experience presenting a poster at a conference:

> When I presented a research poster at a conference, it was equal parts nerve-wracking and confidence-boosting. Accurately communicating what you find, how you found it, and why it is important (whether in a paper or presentation) is one of the hardest parts of research. My biggest worry about the presentation wasn't communicating the methods or findings because I had spent months immersed in the study. It was about explaining why the research matters and why the findings were interesting. Once you get past the planning stages of a study, it is easy to get lost in the tasks and forget why it matters. On top of reminding yourself, you then are faced with conveying that significance to other professionals in the field.

In both of these types of presentations—giving a talk and presenting a poster—it's important to know your audience, as it is with writing papers. At conferences that are more general in nature, you will want to refrain from using jargon, words or expressions that are specific to your field and therefore are difficult to understand, to make your work understandable to a range of individuals. Simplify your language and make sure you know what each term you use means, and give any terms people may not know a brief definition. A good way to test this strategy is to practice your talk or poster presentation in front of friends who are in your discipline as well as those who are not. Get them to tell you if anything is unclear and then try again until they all understand everything but you are not simplifying things so much that they are bored or offended. At a conference that is more focused on your specific field, major, or area of work (e.g., American Economic Association) or a very specific subset of your field (e.g., Economic Science Association), you can assume that people at the conference have a basic understanding of the field. Questions can include clarifications about what you've said ("At which library in Rome did you collect your primary source materials?") or a broader question about why your research is important ("How will this research impact our understanding of global climate change?"). You will sometimes get questions or even challenges to your methodology, findings, or interpretation of results, so make sure you know your own work well enough, as well as other work in your area, to be able to argue for a particular viewpoint or defend your work. Most of the time, people are asking you questions because they are genuinely interested in your work, not because they are trying to trip you up. You should try to anticipate some of the questions people might ask and also ask your advisors and peer researchers to think of questions. You can also think about the weaknesses of your work—no project is flawless—and anticipate the answers that you might give in case someone highlights them. One strategy to take when answering questions is to repeat the gist of the question to give yourself a little time to think and to ensure that you are answering the right question (particularly if the questioner is long winded)—saying "If I understand correctly, you are asking . . ." before giving your answer. Although answering questions posed by others takes practice, you should always remember that you know your work better than anyone else—you are the one who has been putting in the long hours conducting the research—so you are very prepared to answer any questions that may come your way.

Honors Research

A great way to have a positive research experience that ends in a tangible presentation of your work is to complete an honors research project. Although different schools have different requirements for an honors project, in general it involves committing at least a year to an independent research project that culminates in a final product such as a paper, performance, production, or presentation. Honors research is an opportunity to spend a full year or more conducting research on a topic that interests you or that you can learn from. It will give you the opportunity to apply all of the skills you have accumulated from your past academic experiences to a comprehensive project. Even if you have worked on research projects before, an honors project will give you the opportunity to experience the research project from start to finish. Completing an honors project will allow you to work closely with a faculty mentor to build research skills and in some cases collaborate with other students who are just as intellectually curious as you are. As this project will be the culmination of your degree in your department or program, it will allow you to apply the concepts and methods from your college courses to your own research. Working on an independent research project makes you a very attractive candidate to graduate schools (Hauhart & Grahe, 2015), as this experience is close to what you will spend the majority of your time in a research-focused graduate program doing. Even if you do not end up going into an academic career, completing an independent honors project can yield a number of positive benefits, including making you more attractive to potential employers (Hauhart & Grahe, 2015). Independent research can sharpen your problem-solving, reading, writing, and speaking skills, while enabling you to grow as a critical independent thinker (Cabrera, Crissman, Bernal, Nora, Terenzini, & Pascarella, 2002; Cosgrove, 1986; Kardash, 2000). In addition, because you are working closely with an advisor or other faculty member, you will receive a strong recommendation letter from someone who has seen a wide range of your skills over a long time period.

Most honors projects require some end project, usually a research paper or creative work. Often this final work is evaluated by a team of at least 3–4 faculty, so you receive feedback about your research skills as well as your writing or performance. In some departments and schools, faculty will assign a level of honors that denotes the strength of the project. At our institution, the best projects get "highest honors." Receiving this status communicates to others (e.g., graduate schools, employers) that your project is of the highest caliber. Completing a year-long honors project is a very rewarding experience and, if your project is successful, it could also result in a publication or production that is important to your field. At some schools, you may also be able to apply for money to support your project, including money for materials, resources, travel, and salary. Even if you have just begun your first year of college, it is never too early to begin thinking about an honors or senior project. You can start preparing for honors early in your college career by planning to take specific courses and gaining the right research experience to be sure that you are qualified to commit a year to working on your honors project. For example, many honors advisors will require that you take at least one course from them or that you have worked on research with them for a certain period of time. Activity 5.2 will help you learn more about honors research at your school and at other institutions so that you get a sense of what is required and what other scholars have done to earn honors. Keep in mind that there may be GPA requirements that you will need to meet and in some cases application processes that you will need to go through. Knowing what these are beforehand can help you spend time making sure you are able to complete an honors project by the time you get to be a junior or senior.

Articles

If your discipline is a field that disseminates information through journal articles, you can consider presenting your work in a journal once you believe that you have findings that are interesting enough to share with others in the field. Most academic journals publish papers that are reviewed by a small group of individuals who are experts in the field who deem the paper worthy of publication. This type of journal is called "peer reviewed." There are a few ways to tell if a journal is peer reviewed. Some databases will let you search for a specific topic in peer-reviewed journals only. You can also find the official website of the journal, which should tell you if the journal is peer reviewed. After writing your manuscript, you will submit it to one specific journal, where it will go through this peer-review process. This review process can take several weeks or as long as several months, so you

Activity 5.2: Learning about Honors Research at Your School and at Others

A good way to think about an honors project and to see what that project entails is to look for honors projects that were completed by undergraduates at your school. Go to the website for your university and type "honors project" or "honors thesis" into the search engine, and you should be able to sift through the links there to find examples of undergraduate honors projects. If your college has an undergraduate research center or an honors college, you should be able to find that by typing in keywords such as "undergraduate research" or "student research" on the website. Go check out the projects and get an idea of what your peers are doing. Many colleges also publish their students' honors projects online through the library. To check this out, go to your library's website and look at the page that lists the databases—one of the links or databases listed will probably have the word "Theses" in the title. This should give you an option to browse through honors theses and search by keywords just like you would do with other databases. Check out the theses that are in the department in which you are majoring or that interests you the most.

should plan accordingly. In general, the initial review of a manuscript takes approximately 30–60 days in the natural sciences and 30–90 days in the humanities and social sciences. The style and presentation of your manuscript will differ drastically by field and also differ based on the individual journal. How do you choose which journal to publish in? You and your advisor, and any coauthors, will make this decision based on the alignment of your paper with the mission and breadth of the journal. You can do an Internet search for journals in your field, and most disciplines have a listing of journals that publish research in the area. Your librarian will most likely be able to provide you with a list. You'll want to find the goal or mission of the journal on the journal's website and check to see if it matches up with the scope and content of your paper. The mission statement will give you information about the breadth of research it publishes as well as the type of paper (e.g., review article, experimental study, etc.). By choosing a journal you are identifying your audience, which will help you write the manuscript in the appropriate tone. Each journal will have instructions on its website about how to prepare your manuscript, and most will have an online submission process. The site will provide guidelines for submission that are specific to that journal. For example, they will give page limits, citation style, abbreviation standards, and so on. There are several journals out there that focus on publishing undergraduate research. These journals can be a great place to begin the publishing process, as they often have clear guidelines.

Once you submit your article, it will be assigned to an editor at the journal who will make a decision as to whether it is appropriate for that particular journal and good enough to send out to reviewers, who are experts in the field of your particular research topic. These reviewers will perform an evaluation and make their recommendation to the editor as to whether they think it is worthy of being published in that journal. If so, you will receive a response from the editor encouraging you to revise your document, taking into consideration the critiques of the reviewers, in order to improve your manuscript, and then to resubmit it. The editor then chooses whether or not to publish your paper. It may take several revisions for the editor to deem your paper worthy of publication in the journal. If your paper is rejected at any point, you can then submit it to another journal. If you received reviews of your paper, this can be a good opportunity to improve it so it is more likely to be accepted at another journal. Rejections can be extremely frustrating and disheartening—we have been rejected many times!—but you can look at them as opportunities to take in the feedback and improve your paper. The editor may also give you suggestions of other journals where your paper may be a better fit. For example, if you sent your manuscript to a general journal in your field, it may be better suited for a more specialized journal. It can be very difficult to find a home for your particular research, and your advisor or someone else who has had a lot of experience in the field is probably best suited to make that decision. The range of academic journals goes from very broad to extremely narrow in scope. There are journals in which it is extremely difficult to get an article published (i.e., rejection rate over 90%) and others that are more accepting and work with authors to help them develop their papers. Journals that publish only work written by undergraduate students can be good outlets, as they often have lower rejection rates than other journals in a given field. Wherever you decide to submit your manuscript, you should understand who your audience will be—is it broad or more specialized—to determine the content you will present. In some cases, researchers are invited to write journal articles for a specific issue; in others they might be asked to write based on their expertise. We don't have space to talk about all of the tips we have for you for writing a quality journal article—there are whole books dedicated to that! (e.g., Belcher, 2009; Thomson & Kamler, 2013).

Books

In the humanities and in many of the social sciences, books are still the standard for tenure and promotion for faculty (see Chapter 4 for an explanation of tenure and promotion). Scholars may write articles, but most will eventually write a book. The book may develop from the dissertation, which may be motivated by your work as an undergraduate. The book may also be a revision, editing and updating of work over time. The book may be a revision of other work, including a dissertation or journal articles, that have a longer sense of depth, breadth, and reflection than an individual article or set of articles might have had. Books may be a collection of works, be they individually written chapters, poems, or stories. Books undergo the peer review process at both the proposal stage and when the manuscript is complete. We encourage you to start thinking of types of books you'd like to write and working on the plans now in shorter forms, including articles and essays. We also encourage you to consider writing books for the general public. Such books are often called trade books, and they are an important aspect of public scholarship and reading in general. A popular trade book may have a lower price point than an academic book and may become quite popular. Some academics write both scholarly and trade books so that their research reaches a wider audience. Textbooks are often written by scholars or in consult with scholars for use in classes and take comprehensive approaches to a subject. Textbooks are very helpful for those who are starting to learn about a topic or who want an overview. Those who are starting out in research may benefit from reading multiple textbooks on the same or similar topic for different perspectives and for references to read. Textbooks may be written for both knowledge and for profit. Textbooks are often lucrative for the author but expensive for the student reader.

SOCIAL MEDIA FOR AN ACADEMIC AND PROFESSIONAL AUDIENCE

Many academics are using social media to promote their scholarly work and share research ideas. There are many researchers who embrace the idea of using social media for research purposes, while others are more reluctant to do so. Be sure to run any plans for social media posting by your advisor, particularly for collaborative work. Research from Lupton (2014) provides the pros and cons of using social media as part of the writing and presentation process. Examples of pros are networking, including academic, nonacademic, and support networks, and sharing and getting feedback on research. Examples of cons are the negative perception of social media in the academy, including the perception of self-promotion; the obligations and time commitment of social media; and consequences such as plagiarism and the unauthorized commercialization of your work due to lack of copyright protection.

Different types of social media have pros and cons as well. Facebook, Tumblr, Reddit, and Google+ allow for longer interactions between text, video, and comments. Twitter's restricted 140 characters make for a quick read but may obscure the nuance of the message or interactions. Twitter

TEXTBOX 5.1: CONSIDERATIONS AND CAUTIONS WHEN SHARING YOUR WORK

- Put your name on slides and posters.
- Don't hesitate to ask someone to give you credit when sharing your work in any form, including when telling people about it on social media.
- People's work sometimes gets scooped—set up a Google alert about you and your work.
- If you feel your authorship rights have been violated, contact faculty or your campus research office about it.

flare-ups are common in part because of misunderstandings and an inability to build rapport. Official websites are good for letting people know that you are legitimately affiliated with your college or university; they allow you to integrate that information with other social media. But if not regularly updated, websites can become outdated. Blogs are easy to update, and sites including Wordpress and Blogger have the utility of websites. Textbox 5.1 contains things to think about when you're sharing your research work in any form.

CURRICULA VITAE AND RESUMES

Once you have presented your research at a conference, written a paper, or contributed to or written a book, you will want to share these experiences and accomplishments with relevant audiences to showcase your academic and personal success and skills. Two particularly useful forms of sharing your accomplishments are the resume and the CV (curriculum vitae—Latin for the course of [one's] life). A resume is typically used for applying for jobs, internships, or other such positions. The format for a resume is usually a one-page document that highlights your education, work experience, and skills. It summarizes your qualifications for the position to which you are applying. Your resume should contain a statement that describes the position for which you are applying and information about education, jobs you've had, organizations, leadership experiences, relevant class projects, and skills learned in and out of the classroom. A CV is an academic version of a resume. It is typically used when applying for research grants, graduate school, and academic positions. Unlike a resume, your CV can be more than one page in length. You should have a section about your professional experiences and research presentations such as conference posters and journal articles, as well as a section listing any awards, honors, or grants you have received. You may include other sections based on who will read this, including relevant skills, membership in professional organizations like an academic society, foreign language proficiencies, training, etc. Finally, provide a list of references (at least three) who can speak about your academic accomplishments and your academic potential. There are some good resources giving additional tips on writing a quality CV as well as examples of CVs linked on our website.

CONCLUSION: YOU CAN DO IT!

With your newfound knowledge of writing and presenting, you are now well on your way to a complete research experience. Remember, work or research from an undergraduate class or research project may be acceptable for presentation or publication! You may already have something that is good for the world, so don't be daunted into thinking that presenting and publishing is not for you. In some disciplines, you work with professors on topics that are much more quickly and readily publishable. In others the research takes more time, so we encourage you to share your work in progress at student conferences and in publications.

In our next chapter we address the specific challenges that underrepresented scholars may face. It is important to read the chapter so that you can best support all of your fellow scholars throughout the research process.

CHAPTER 6

Underrepresented Scholars in the Academy

Making a Way

> But this is true: A university is a human invention for the transmission of knowledge and culture from generation to generation, through the training of quick minds and pure hearts, and for this work no other human invention will suffice, not even trade and industrial schools.
>
> —"The Talented Tenth" by W. E. B. DuBois

Your attendance at college is quite possibly someone's dream manifested. That dream may be your own, or that of individuals who have supported you along the way—teachers, family, and friends. Their lives will be changed through your experiences. Many of our own undergraduate research scholars are the first in their families to go to college. Their entire families go to college with them through the knowledge that students bring from their homes and their communities, and the information and opportunities that they bring back home. There is an amazing blend of legacy students and new spirits, from backgrounds or families previously excluded from higher education, that results in new experiences and new ideas on college campuses. The research these students do enhances those experiences.

In June 2016, the Supreme Court of the United States (SCOTUS) upheld that race can be used as a consideration in college admissions. This decision will have an impact on how campuses include students in the future. Scholar Ebony Lambert describes how we can use the decision to spur changes in how race is approached on university campuses:

> The SCOTUS decision has very tangible implications for the way we should discuss race on college campuses—and in a way sets a clear course of action for current and future scholars and educators. By upholding "race-conscious admissions" procedures, the SCOTUS provided a foundation for educational institutions to discuss injustice. Here the decision accounts for previous structural and institutional injustices

> (and those that persist today), and leaves room for race as a factor in admissions to be cited as a method by which we can begin to eradicate these injustices. The case in and of itself makes clear the need for campus and curricular conversation regarding race, how systems of oppression continue to exist, operate, and marginalize, and finally the ways in which the academy is responsible for advocating and fostering socially just practices and atmospheres for students, educators, and communities of color.

Underrepresented scholars are crucial for the future of research, as we have highlighted in the section in Chapter 2 entitled "Research by Underrepresented Scholars," and as we will demonstrate throughout this chapter. Yet sometimes, as the old saying based on Isaiah 43:19 in the Christian Bible suggests, we have to "make a way where there is no way." That's because many of us will enter areas of research that are new to us and to our communities. There may not be many people from our particular social and cultural backgrounds who have participated in undergraduate research. We became involved in the support of underrepresented scholars at the College of William & Mary, because, at our school, there weren't as many underrepresented students doing undergraduate research as students from represented groups, and we wanted to find out why and what we could do about it. We found out that we could do a lot. And you, the reader, can, too. The questions in Textbox 6.1 ask you to connect your identity to your research experiences, as the two are inextricably intertwined.

We created the William & Mary Scholars Undergraduate Research Experience (WMSURE) to support students in learning how to do research and to allow our scholars to share in a cohort or group experience, with opportunities to get to know and support each other academically and socially. WMSURE provides weekly workshops, ongoing advising, and financial aid to help students succeed in undergraduate research. WMSURE workshops are open to all William & Mary students and to the public, which we believe helps our program achieve the broader goals of providing resources for success to all students, and educating *everyone* about issues that are particularly important to scholars from underrepresented groups.

Scholar DaVon Maddox illustrates how this program has allowed students to tap into and learn more about their own identities through research:

> Racial discrimination is the basis behind my involvement in research. On campus, I feel as an outcast to my race because I am involved in organizations that are stereotypically seen as being reserved for the racial minority. I don't seclude myself to the minority group in hopes of feeling like I belong to a misunderstood or underrepresented group. As

TEXTBOX 6.1: CONNECTING THE PERSONAL TO THE INSTITUTIONAL

- Who are you?
- How do you define/describe your background/culture/heritage?
- How do you define yourself as an individual?
- What are the particular lenses and perspectives that you bring to your research because of your identity/background/culture/heritage?
- What are the particular lenses and perspectives that you bring to your research from your individual experiences?

a result, the prejudice is the driving force behind my research involvement. I am a part of a research lab that primarily looks at prejudices and stereotypes. Now, the discrimination and injustices are my fuel to project further into the research studies, with hopes of finding answers to underlying questions in my personal experience.

Based on what we've learned from our WMSURE experiences, this chapter is devoted to helping students who are underrepresented be prepared for some of the particular experiences they may face as undergraduate researchers. The chapter is also designed to inform your more represented peers about the experiences of underrepresented scholars so that they can be supportive peers as undergraduate scholars, graduate students, and future faculty. We want you to start with who you are: an individual with a culture who is part of one or more communities. As researchers, many of you will enter and join other communities. Your identities may change as you develop and grow as a person and a researcher, but who you are and who you want to be remain most important.

Diversity and inclusion in research is important across many areas and disciplines, as researchers from different backgrounds, cultures, and experiences have different ways of looking at the world. The differences in these focuses and lenses are crucial to the progression of inquiry. The continued integration and inclusion of underrepresented scholars is very important because it contributes to equity in education—ensuring that *all* students not only access but thrive in academic opportunities (also see a more detailed definition of equity below from the Association of American Colleges and Universities [AAC&U]). While desegregation brought about legal changes to pave the way for students from underrepresented backgrounds to attend a wider variety of colleges and universities, full integration has not occurred (Horsford, 2011). Most larger, wealthier universities, in particular, are highly segregated and have an even less diverse faculty than student body. When

Textbox 6.2:
Questions to Consider About Integration in Higher Education

- When were colleges and universities in your area integrated by law? By actual student enrollment?
- What is the percentage of first-generation status students? International students? How does this demographic information compare to information about faculty?

universities are diverse, the diversity tends to be mostly in one racial group or category. For example, a university might advertise that one-third of its students are students of color, but most of these students may be from the same background while students from other backgrounds continue to be underrepresented. The questions in Textbox 6.2 help you ask questions about integration at your school.

Historically Black colleges and universities (HBCUs), minority-serving institutions (MSIs), and women's colleges have existed to serve populations that were barred from what we now view as our highly research-active private and our flagship public predominately White institutions (PWIs) and universities. HBCUs and MSIs are vital, as they serve populations and create spaces for research scholars that are not present in the same numbers at PWIs. Even if the student population is more balanced today at a university, the faculty and administrative populations may not be. The persistence of HBCUs, MSIs, and women's colleges reminds us of the challenges that continue on campuses due to structural injustices in education and in society. Programs including the National Science Foundation–sponsored HBCU-UP continue to support research by scholars at HBCUs in recognition of the large number of African American researchers that are educated at HBCUs. Professor Rhonda Fitzgerald describes her experiences learning and teaching at HBCUs:

> It was finally over *and* just beginning. After a total of 8 years of graduate studies by way of Hampton University and Virginia Commonwealth University, respectively, I was now Rhonda D. Ellis, PhD. I was now in a position to join the ranks of those whom I've been in awe of for my entire adult life. I would commence this new journey as a tenure-track professor.
>
> As a biostatistician, I considered pursuing professorships at a top research university. This was the road traveled by many of my fellow graduates. Surely, I would thrive in a "publish or perish" atmosphere. Yes! Success awaited me at UCLA, Duke, or even Columbia.

However, there was something tugging at my conscience. While I was the proud owner of a newly minted doctorate from VCU's Medical College of Virginia Campus, it was Hampton University that served as the initial incubator for my burgeoning intellect. To be honest, I owed this HBCU just as much credit, if not more, for my success as I did VCU. It was clear: I had a debt to repay.

You see, there were still many HBCUs that needed people like me—young, black STEM PhDs. Why? There is a notable lack of African Americans pursing STEM disciplines. This is, in part, due to the lack of African American STEM instructors on the university level. It's an unfortunate self-perpetuating cycle. One way to help break the cycle is for people like me to make teaching at an HBCU our first choice.

Needless to say, I'm in love with my first choice. In fact, with support from organizations like the Quality Education for Minorities Network, I've geared my latest National Science Foundation HBCU-UP research grant toward better developing STEM majors at my HBCU. I am still thrilled that my "beginning" at an HBCU has contributed to the beginning of more African American representation in STEM fields.

For many students who arrive at campuses that don't reflect their backgrounds, the differences can be visceral. James Ryan (2010) describes the inequalities in experiences in high schools in Richmond, Virginia, just 45 minutes from the College of William & Mary, where this book was written, and so these inequalities impact our students, our own classrooms, and research findings. Differences across college campuses are visceral as well. The resources afforded to White research universities are evident, while many colleges and universities that serve underrepresented populations need increased financial support that would make research possible for more students.

INSTITUTIONAL CONTEXT

Diversity and inclusion are critical for research because they allow for more multifaceted thinking and perspectives. Diversity in scholarship allows us to grasp a fuller understanding of humanity and culture. It also allows for a more comprehensive approach to science. As Medin and Lee (2012) state:

> [A]ttention to cultural membership and cultural practices is central to equity goals and national needs, but also equally important for the construction of knowledge and for the enterprise of science itself. Moreover, we cannot and do not shed our cultural practices at the door when we enter the domain of science, science education, or science learning.

The Council on Undergraduate Research (CUR) expands on this approach and explains that, if underrepresented scholars aren't included in undergraduate research, then we can't do what the CUR calls "expanding frontiers of knowledge." Deficit and other negative approaches to research have left communities and individuals, including many of us scholars, wary about the methods and intentions of some research.

Sociology professor and researcher Tukufu Zuberi (2000) exemplifies such issues through his work. His research demonstrates that "racial statistics lie" through "misused statistical methods" (p. xxii), and he challenges the notion that empirical research, or research based on actual observation, is free from biases. Zuberi (2000) shows that the methods that are used to draw conclusions about race are not protected in the empiricism. The root of the methodology still lies with the researcher and assumes the researcher's bias. He challenges how bias has led researchers to even create race and gender, which are now widely viewed by social scientists as a continuum of social constructs. Zuberi gives the history of how bias, and racial bias in particular, came to be in our empirical models of research and explains the greater social implications of this bias through examples of how science has been used to support racism and eugenics, among other social ills.

Organizations That Support Diversity and Inclusion in Research

Association of American Colleges and Universities (AAC&U). The AAC&U addresses inclusion in research through its focus on diversity and provides useful definitions for the higher education context (AAC&U, n.d.). For example, the AAC&U defines equity as "the creation of opportunities for historically underrepresented populations to have equal access to and participate in educational programs that are capable of closing the achievement gaps in student success and completion" (AAC&U, n.d., para. 7).

Federal and private foundations also value inclusion and have set up criteria and models for inclusion that also start at the undergraduate level. Their arguments form a good baseline for thinking about why it is important for you to read and put the information in this chapter into action so that you can appreciate the full value of your work to the academy if you are an underrepresented scholar and, if you are not underrepresented, so that you can support and help to prepare for the more inclusive academy that we are working together to build.

National Science Foundation (NSF). All NSF proposals must include an explanation of the "broader impacts" of the proposed activity(NSF, 2016). Under this initiative, the NSF requires an explanation of how the activities associated with research broaden the participation of underrepresented groups (e.g., gender, ethnicity, accessibility, and geographic area).

National Endowment for the Humanities (NEH). Such sentiments and mandates are not confined just to natural and social science research, but are also found in the arts to show commitment to and the importance of diversity and inclusion across disciplines. The NEH includes a comprehensive impacts aspect (NEH, n.d.). Its activities are crucial for helping cultures to flourish. The NEH offers specific awards for faculty at historically Black colleges and universities and Hispanic serving institutions, and awards funds for Sustaining Cultural Heritage Collections programs and Documenting Endangered Languages programs. NEH Impact Reports describe NEH-supported initiatives for underrepresented populations.

The Ford Foundation. The Ford Foundation seeks "to increase the diversity of the nation's college and university faculties by increasing their ethnic and racial diversity, to maximize the educational benefits of diversity, and to increase the number of professors who can and will use diversity as a resource for enriching the education of all students." The Ford Pre-Doctoral Fellowship in conjunction with the National Academies of Sciences, Engineering, and Medicine (2016a, b) awards approximately 60 predoctoral fellowships. These fellowships provide 3 years of support for individuals engaged in graduate study leading to a Doctor of Philosophy (PhD) or Doctor of Science (ScD) degree.

The Mellon Foundation. The Mellon Foundation contends that "Building just and durable democracies in the 21st century depends on extending the benefits of higher education to all, and enabling students of all social, ethnic, and economic backgrounds to experience and value diversity and inclusiveness in their pursuit of learning."

Some programs receive federal or state funding and support organizations at particular universities. Two prominent examples of such programs are the McNair Scholars Program and the Mellon Mays Undergraduate Fellowship Program, which focus on increasing the diversity of PhD students and faculty.

Undergraduate Equity-Focused Organizations

As described by the National Science Foundation, "The Research Experiences for Undergraduates (REU) program supports active research participation by undergraduate students in any of the areas of research funded by the National Science Foundation." REU projects involve students in meaningful ways in ongoing research programs or in research projects specifically designed for the REU program.

The foundations and organizations described above illustrate just a few of the ways that research organizations have conceptualized the value of diversity in research and in the academy. All of the programs we describe take a strong stand on the value of diversity and inclusion. Similarly, in our research we take a social justice approach to our stance on diversity and inclusion. We believe that scholars from all backgrounds deserve to be researchers at colleges and universities, yet people from our backgrounds have been denied the opportunity in the past. So we wrote this book to make a way for scholars in the future. Social justice research, further explored in Chapter 7, gives us a framework for collaborating with and in communities, so that the differences between researcher and participant are mitigated and inequalities are diminished.

FACING CHALLENGES

All students, on every level of education, should have questions about the intent of their education and their educators. Research scholars should actively seek out answers to those questions. We want to help you think critically about the challenges that underrepresented scholars may face due to discrimination, particularly due to racism. We present theoretical, qualitative, and quantitative models for each topic. We explain how such issues have resulted in stereotype threat and solo status. Most important, we give you strategies to address your experiences and feelings by acknowledging what you are feeling and by seeking support—using your experiences and research for positive change. We put a heavy emphasis on trust, and on helping you seek out peers, faculty, and others in the university setting that you can trust to have your best interests in mind and at heart.

In any learning situation, students have questions about the safety of their environment. Three key questions that arise are: Am I safe here? Will I be successful here? Is there something in this for me?

Safety may be defined as the physical safety (on a campus in general, in a research lab, or in a particular learning setting) and the social and emotional safety that being a student in a class or a member of a research group of scholars should bring. We address how to talk about such issues to gauge safety and to ensure a safe environment.

The question "Will I be successful?" is particularly challenging in the research context. It is difficult to answer for all scholars because it's harder to predict whether or not research will be successful; it is often a less controlled environment than, say, a class. Add then a situation where you are underrepresented—or where there are fewer people from a particular background who are role models who can show you the way—and the risk management of success becomes even more difficult.

Along these lines, the question "Is there something in it for me?" arises because the topics that underrepresented students are interested in might not be represented in the curriculum and in research opportunities at every college and university. We show you how to make do with what you have and how to find other scholars and faculty who share your research interests.

In addition, Maslow's hierarchy of needs (1943) states that physiological needs undergird safety, love, esteem, and self-actualization. Newer research indicates that Maslow's model may not actually be a hierarchy, and that love, esteem, and self-actualization are more integrally intertwined. Critics have claimed that Maslow's observations are based on a dominant U.S. society model, which privileges the individual, rather than on societies that are more collective in nature, such as those societies of individuals from underrepresented ethnic and racial communities that may tend to be more community-oriented (Triandis, Bontempo, & Villareal, 1988). We further explain this concept by providing you with definitions of discrimination that focus on you as a scholar and your scholarly experiences. We want to help you be aware of how discrimination manifests in research contexts. We also want to give you ways to address it as it occurs.

Discrimination

> What we can see depends heavily on what our culture has trained us to look for. (Nell Irvin Painter, 2010, p. 16)

To understand the various facets of discrimination and how they apply in research contexts—so that you can name your experience and share it with others—it is important to have a well-defined understanding of the types of discrimination that you may have experienced before college and in college so far. We then focus right away on what you, as a research scholar, can do about them.

Personally Mediated Discrimination

Personally mediated discrimination is characterized by prejudicial attitudes involving differential expectations about capabilities, motives, and intent of others according to race (racism), gender identification (particularly misogyny, which is directed at women), sexual orientation (particularly homophobia and transphobia, which are directed at LGBTIQ people), and/or class (classism). Discrimination based in racism involves the differential actions and behaviors toward others according to their race. Such racism also includes stereotyping, particularly when individuals disrespect, are suspicious of, devalue, and dehumanize others (see Jones, 1997).

Personally mediated discrimination in the form of microaggressions on college and university campuses has been a recent focus in both academic research and popular media. The term *microaggression* refers to everyday biases and indignities faced by members of marginalized groups (Sue, 2010). For instance, the statement "I've never heard an intelligent person talk the way you do" is a common microaggression. Another microaggression, which tends to be made toward African Americans, is the question "Are you here on scholarship?" Microaggressions can imply that members of underrepresented racial or ethnic groups are not expected to be at college, and if they are, they are supposed to be doing certain things like athletics or should be grouped in certain subject areas or perform in a certain way. Every few years, researchers who believe in the difference in abilities of women, African Americans, or people with lower incomes make a splash with a study that states that one racial, ethnic, or social group is statistically less intelligent or verbose based on IQ or vocabulary assessments (Hart & Risley, 1995; Hernstein & Murray, 1994). Then there are scholars who comprehensively counter those arguments with their own work, including Hilliard (1990) and Charity Hudley and Mallinson (2010). But more often, personally mediated discrimination happens on the local level.

As described in the collection Presumed Incompetent, edited by scholars and professors Gabriella Gutiérrez y Muhs, Yolanda Flores Niemann, Carmen G. González, and Angela P. Harris (2012), discrimination can also be framed in a research risk-management model. Such a model states that people are wary of accepting students they don't think will succeed, and like their research spaces to be familiar and safe, as many faculty may not have been educated about how to teach in multicultural educational spaces, and thus are less comfortable with students from different sociocultural backgrounds. Therefore, it is important to find at least some research experience that has and values diversity. Sometimes nondiverse labs are unavoidable, but ask around for projects and opportunities that may be outside of a specific research area or outside of your university so that you gain the social and cultural competencies that should come with a research experience.

Personally mediated discrimination can be the result of both explicit and implicit bias. Explicit bias refers to prejudicial attitudes that individuals hold that they are aware of. Implicit bias refers to when the individual is unaware of their prejudicial attitudes or stereotypes. In this case, although an individual may feel that they consider all groups equal and that they don't discriminate against individuals of one group, they do hold unconscious prejudicial attitudes. This form of bias can manifest itself as more subtle, but just as harmful, discrimination. For example, psychology researchers demonstrated that both male and female science professors from research-intensive universities rated female students who applied for a lab position as less competent and hirable than male students; more negative evaluations were associated with the amount of implicit bias they held (Moss-Racusin, Dovidio, Brescoll, Graham, & Handelsman, 2012).

What to do About Personally Mediated Discrimination. *The Black Academic's Guide to Winning Tenure Without Losing Your Soul* (Rockquemore & Laszloffy, 2008) is written for faculty, but is good for readers at any point on their academic path, especially those who have decided how to pick their causes and battles given the power dynamics of academic relationships. Rockquemore and Laszloffy note that the important thing to consider is that they are your battles—what you choose to challenge may differ from what faculty, other students, or your family would choose. Chart your own course as you approach discrimination, but be aware of the history and outcomes of others. Decide which issues you are going to take on, while acknowledging those you don't. It's okay. Every experience is different.

If you are in a situation where you are or might be experiencing discrimination and/or harassment, document everything and let someone at your university know, even if you don't file a formal complaint. Documenting your experiences can help your university see patterns when you might think things are happening only to you. At many schools it is the dean or dean of students' office that will help you with the process. Find a faculty member or administrator that you feel comfortable with and let them help you go from there. You may want to report to someone who is outside of your institution. You may also turn for support to those with experience in these areas, including faculty who are affiliated with the Ford Foundation, the Mellon Foundation, and the McNair Scholarship Program, as well as other institutional programs, such as WMSURE.

Institutional Discrimination

Institutional discrimination encompasses the ways in which organizations, institutions, governments, and other entities institute and use practices and policies that allow for differential treatment of certain members of a society (Jones, C. P., 2002; Jones, J. M., 1997). Such policies can be de jure, as in the case of apartheid in South Africa and Jim Crow laws in the American South, or de facto, as in the case of current school segregation in the United States. Segregated housing policies and patterns, discriminatory employment and promotion policies, racial profiling, inequities in health care, segregated churches, and educational curricula that ignore and distort the history of underrepresented groups at colleges and universities and in research are all part of institutionalized discrimination. The English-Only Movement, which seeks to make English the official language of the United States, is a current push for institutionalized linguistic racism (Schmidt, 2002). Educational institutional racism is at the root of why many students have not had access to higher education. Even after colleges and universities desegregated, admissions and completion rates were slow to grow. Institutional racism in undergraduate research is also tied to the social networks that faculty use to find researchers and students and to their willingness to mitigate risk and preserve ties by providing fewer opportunities for underrepresented students.

As scholar Ebony Lambert (2016) explains in her senior honors thesis:

> Further examples of institutional racism and its reach on the college level include "culturally biased curricula, lack of people of color in positions of power, lack of institutional support, and passive responses to individual racist acts" (Cokley, 2006). It can also be seen at the university level when buildings are named disproportionately after White Americans and/or people who have advocated for or created means by which people of color could be further marginalized, in the number of minority faculty that are tenured, in the admissions process (the weighing of standardized test scores), and also in disparities in graduation rates and academic prosperity across races or ethnicities (an "achievement/opportunity" gap). Recent examples of institutional racism that have gotten news coverage include the Stokes (2013) spoken word video regarding the struggles and frustrations of African American males at UCLA, as well as in the call to action demanding change put out by the Black Student Alliance at the University of Virginia (Seal, 2015).

As a result of exposure to institutional discrimination at the collegiate level, some students from underrepresented backgrounds report feeling that their appearance (race/skin complexion, hair style or texture, style of dress, etc.) affects the way they are treated by educators both in and outside the classroom (i.e., in office hours, academic advising, at extracurricular events, etc.). Additionally, some underrepresented scholars report feeling that their race negatively influences the way their professors and peers perceive them and their academic abilities, and some even go as far as to say that they had altered their hair or style of dress on various occasions in order to avoid negatively representing themselves or their race. Here students spoke about such experiences as not getting specific hairstyles or hair weaves and/or refraining from wearing sweatpants or athletic gear to class in order to avoid being perceived as either "ghetto" or lazy. Analysis of our survey data about the impact of the WMSURE program also supported this research, such that students from underrepresented backgrounds reported experiencing more stress and concern about how they will be perceived or treated due to their sociocultural backgrounds, as well as more everyday discrimination.

Scholars from underrepresented backgrounds also cite a general sense of unfamiliarity with their culture as a factor that contributes to their experiences of stereotype threat and impostor syndrome (see sections below). To these students, their professors and fellow peers seem poorly educated on their sociocultural history and realities, and this disconnect may lead them to question whether they were actually wanted or welcome at their educational institution.

What to do About Institutional Discrimination. First, it is important to recognize that you have worth both as a person and as a scholar, and to recognize that the lack of sociocultural representation or support you may experience at your university is not indicative of your potential success or whether or not you belong in academia or in your particular field.

It is also important to make others aware, particularly peers and faculty, when you face institutional discrimination. We all have inherited an education system in the United States that prevents many students from seeing their true worth. We can, however, be brave and confront racism with a restorative justice approach. According to the International Institute for Restorative Practices, "restorative practices is a social science that studies how to build social capital and achieve social discipline through participatory learning and decision-making. . . .The aim of restorative practices is to develop community and to manage conflict and tensions by repairing harm and building relationship" (Wachtel, 2013, pp. 1, 4).

Such an approach is challenging, but it allows for both the accuser and the accused to come to mutual common ground; it differs from the more traditional punitive approach that may lead to justice but no resolution. Such approaches are important when researchers must continue to work together in the short or long term.

Undergraduates have been crucial to mitigating and ending institutional discrimination, from integrating colleges to advocating for new degree programs, including Africana Studies, Asian American and Pacific Islander Studies, Chicanx and Latinx Studies, Ethnic Studies, Native American Studies, and Gender, Sexuality, and Women's Studies. Organizations that brought together students, faculty, and community members have been key historically, and even now with the Black Lives Matter movement. Talk with your research team, no matter what subject you are researching, about institutional discrimination and how it impacts your research experience. Questions to ask include: In the undergraduate research context, are research opportunities and supports distributed evenly across students?

Additionally, based on our own research and experiences, student activism can be very beneficial to scholars when it is balanced with time management and self-care. Student activism can aid not only underrepresented students and students who have experienced personal or institutional discrimination, but also student populations from more represented backgrounds who wish to serve as allies and change-makers. On your own campus, getting involved with cultural, research, or institutional strategic planning groups that involve students can be a great way not only to find/provide language and coping skills for discriminatory experiences for yourself and other students, but also to use scholarship and community-based research approaches to foster social and cultural shifts toward diversity and inclusion on your campus. Thus your research can foster activism and your activism can foster research.

Most campuses have Equal Employment Opportunity (EEO) offices. A growing number of campuses have chief diversity officers or diversity committees that engage with issues of institutional discrimination on their campuses and in higher education. You may also want to bring up issues and challenges you are facing with deans and directors of research; outside agencies including the U.S. Department of Education Office (DOE) for Civil Rights; and academic organizations for your research area or discipline and for higher education in general, such as the AAC&U and CUR.

The DOE Office for Civil Rights provides resources on how to file a discrimination complaint and what your rights are when you do so. The DOE website describes the process for filing a complaint online at www2.ed.gov/about/offices/list/ocr/complaintprocess.html.

Internalized Discrimination

Internalized discrimination is the acceptance of negative ideas about one's own group and its value and worth, which is particularly characterized by not believing in the worth of one's culture, intelligence, or practices (Jones, 1997). In his book *Pedagogy of the Oppressed*, Freire (1970) explains internalized discrimination as follows: "So often do they hear that they are good for nothing, know nothing and are incapable of learning anything—that they are sick, lazy, and unproductive—that in the end they become convinced of their own unfitness" (p. 63). Internal devaluing may pertain to aesthetic, linguistic, intellectual, or other forms of culture. Internalized discrimination has manifested in models of academic insecurity as well as the refusal to speak or value stigmatized languages or linguistics varieties—that is, language varieties subject to prejudice, as Lippi-Green (2012) explains. Internalized discrimination may also manifest as a hesitancy to speak at all because speakers recognize that they speak a stigmatized language variety and think that they are somewhat intellectually inferior or broken. Internalized discrimination in the research context often manifests as self-doubt, as scholars may doubt their abilities as part of the way that they see themselves. It can also lead to the imposter syndrome, which we will explore in one of the following sections.

What to do About Internalized Discrimination. Because internalized discrimination is much harder to face with external action, the rest of the chapter is devoted to approaches that center on being aware of types of feelings you may find yourself having that are related to internalized discrimination and how you can process those feelings with reflection and then action.

Stereotype Threat

Stereotype threat is "[t]he threat of being viewed through the lens of a negative stereotype or the fear of doing something that would inadvertently confirm that stereotype . . ." (Steele, 1999). Stereotype threat can affect any group, but the threat must be relevant to self. Psychologist researcher Claude Steele (2010) frames stereotype threat most saliently in the following passage:

> We came to think of this pressure [stereotype threat] as a "predicament" of identity. An American White woman in an advanced college math class knows at some level that she could be seen as limited because she is a woman; a Black student knows the same thing in almost any challenging academic setting; and a white elite sprinter knows it, too, as he reaches the last 10 meters of a 100-meter race. These people know their group identity. They know how their society views it. They know, at some level, that they are in a predicament: their performance could confirm a bad view of their group and of themselves, as members of their group. . . . This term [stereotype threat] captured the idea of a situational predicament as a contingency of their group identity, a real threat of judgment or treatment in the person's environment that went beyond any limitations within. (pp. 59–60)

The threat also must be variable across different groups and situations. It's important to note that someone doesn't need to believe a stereotype to be impacted by a stereotype; they just need to know it exists. In fact, when a student, particularly a high-achieving student—exactly the type who would be most likely to want to do research or read this book—tries to *disprove* a stereotype, their performance decreases. Steele has spent decades researching stereotype threat and has explained, "*The most achievement oriented students, who were also the most skilled, motivated, and confident,*

TEXTBOX 6.3: QUESTIONS FOR REFLECTION AND WRITING

- What do you value about yourself?
- What is an example of a time when you showed integrity?
- What do others value in you?
- How do others articulate your worth to you?

were the most impaired by stereotype threat" (Steele, 1999, p. 48). That's because better students identify more with school and thus try harder and see more at stake with academic tasks. Stereotype threat can result in distraction, self-consciousness, evaluation apprehension, test anxiety, and loss of motivation.

Stereotype threat can be alleviated through self-affirmation of self-worth and integrity. Females and African American scholars who wrote about a valued characteristic, such as dedication to friends and family, performed better on a subsequent math test. Another study showed that two 15-minute writing exercises at the beginning of a college physics class improved females' performance, especially for those who endorsed the stereotype that women are not as adept at science as men (Cohen, Garcia, Apfel, & Master, 2006; Kost-Smith, Pollock, & Finkelstein, 2010; Martens, Johns, Greenberg, & Schimel, 2006). Textbox 6.3 is designed to help you do self-affirmation writing similar to that which worked in the studies above.

Are such exercises magic bullets? Of course not. But what they represent is the framing of social adversity in college as a shared and short-lived event. Such exercises allow you to attribute adversity to the real sources—and to recognize common and transient aspects of your experiences as adjustment to college, not as fixed unique deficits. Such a change in thinking during freshman year improved GPAs of students, particularly of African American students (Walton & Cohen, 2011).

Research also shows that having researchers with backgrounds that are similar to yours as role models improves your research and scholarly experience (Dasgupta, 2011). For example, females' attitudes, identification, self-efficacy, and career interest in STEM (especially implicit) are enhanced by having a female STEM mentor. African American students showed enhanced self-efficacy and performance after learning about a high-achieving African American scholar. Role models are most effective when they are perceived as similar to you. Professors and guest speakers may serve as effective role models, but exposure can also be virtual (Dasgupta, 2011). Programs that allow you to call and see people online are a good place to start if travel is prohibitive.

Imposter Syndrome

The *imposter syndrome* is the feeling that you don't belong at college, or aren't qualified to do research at all. Perhaps you feel like you got into college because of your race or gender or because the college needed low-income students. Maybe you thought it was just luck or a mistake. People that you like or dislike may even have said that these are the actual reasons you were selected or won't succeed at college, which further contributes to imposter syndrome and relates this phenomenon directly to discrimination.

Imposter syndrome can occur if you do not believe, due to internalized discrimination, that you are as intellectually capable as your peers or have the skills necessary to fulfill the requirements of your role as a student. University of Illinois Urbana-Champaign Counseling Center materials state that these beliefs may lead you to dismiss any academic or career-related successes as based upon external factors such as beginner's luck, extra work effort, networking with influential people, or filling a perceived quota (e.g., "I was only offered the research position because they needed more underrepresented students in the lab"). See more on online at counselingcenter.illinois.edu/brochures/coping-race-related-stress.

As a high-achieving student, you may experience imposter syndrome, which is associated with high-achieving people, as well as with people with high levels of self-consciousness and self-monitoring and people who are concerned with impression-management. Discrimination, subjective evaluations, and underrepresentation are other factors that exacerbate imposter syndrome. Personally mediated and institutional discrimination contribute to fears associated with imposter syndrome. First-generation college students, scholars, and professionals often feel that they are frauds who don't deserve to be charting a new path because their experiences of implicit and explicit biases support this fear. Imposter syndrome can also be exacerbated by subjective reviews. For example, people in creative careers (artists, writers, designers, professors) whose work is reviewed by subjective audiences sometimes feel good only when they get good reviews. People who are in situations where they are severely underrepresented often feel the pressure of needing to represent everyone from their background to show that they and others like them can actually do the research and succeed.

In college, you may be more likely to feel like an imposter because while in high school your teachers may have told you explicitly, "You will succeed," but in college, faculty's expectations for your success may be more implied. In the research context, the imposter syndrome may lead scholars to apply low effort (adopt an ability-avoid goal), maintain a low profile (to shield them from scrutiny), deflect attention away from those areas in which they feel fraudulent, engage in acts of self-sabotage (show up late or unprepared, put in less than a full effort because they believe it won't matter anyway), or even refrain from engaging in research opportunities in the first place. In the research context, because answers are not known, it may feel like you are not the one able to find them out. It may seem that someone more qualified should do the research—but we want to encourage you to think, "Why not me now or me someday!"

How to Address Imposter Syndrome in the Undergraduate Research Context. It is important to acknowledge that you feel like an imposter so that you can talk to someone you trust and seek counseling support. Talking with others will help you separate truth (reality) from what is not true in a given situation and help you to make sure you are living up to your own expectations for academic and research success. Such direct confrontation is important in a research context because it will help you continue to stand toe-to-toe with the faces of uncertainty, doubt, and potential failure.

Solo Status

Solo status is the experience of being the only member of one's race, gender, or culture present in a group. Majority group members often view solos with increased scrutiny and bias (e.g., Crocker & McGraw, 1984). Indeed, even the awareness of being the sole member of one's group is burdensome and has detrimental effects on performance, but does not depend on a stereotype being salient. Experimental research comparing groups' performances on academic tasks (Inzlicht & Ben-Zeev, 2003; Saenz, 1994; Sekaquaptewa & Thompson, 2002; Vohs, Ciarocco, & Baumeister, 2005) has shown that solo status can disrupt cognitive functioning and compromise learning and performance. Students often experience solo status in classes.

Intellectual solo status is also structural and can happen when research from people from your background is not represented or valued. It can also happen when you are the only person with a particular viewpoint. Carter (2012) explains solo status in a quote from a college biology student he interviewed for his study: "For example, my professor in the neuroscience department did not know who Dr. Ben Carson, an African American neurosurgeon at Johns Hopkins Hospital, was." When solo status is related to microagressions, even wanting to go into office hours can be tough. As another student cited by Carter (2012) relates: "Many of my professors were very condescending. On multiple occasions I would attend a professor's office hours and ask for help with understanding problems I got wrong on an exam. The professor would say things like, 'I can't believe you got that wrong. It was easy, but you were only one of six students out of the whole class to get this hard question right. I don't understand how you could get the hard question and not the easy one.'"

Discrimination and solo status are often the reasons many students we work with report that they are still intimidated by office hours. They explain that they are trying to find signs as to which professors will help them feel safe, feel as if they will be successful, think they deserve to be in college and in a research experience, and will help them on their journey.

What to do About Solo Status. It is important to share your experiences and your solo status with faculty, who may or may not be aware of it. Many faculty members have reported to us that they know that there is an issue when there is only one student from a particular background in their research group but they didn't know about the concept of solo status. Moreover, students have shared with us some of their positive experiences with faculty who were upfront about their goals for inclusive classrooms and research spaces. Talk to your fellow students to help you find faculty who are aware of discrimination, imposter syndrome, and solo status. Cohorts that reach across research areas mitigate solo status and will expose you to new ideas. If there are research groups to join or other faculty or students on campus that you can meet with, it will help. You may also ask faculty to help you find researchers from your background in your research area at other universities so that you can learn about their work and experiences even if you can't meet them in person.

Health Effects (by Scholar Ebony Lambert)

Discrimination can have psychological health effects on students, as we discussed earlier in the sections on stereotype threat, imposter syndrome, and internalized discrimination. It is important to note here, however, that these psychological constructs can also lead to very real physiological health effects, both inside and outside of university doors. In conceptualizing links between discrimination and health, Ziersch, Gallaher, Baum, and Bentley (2011) conducted 153 interviews with aboriginal people living in Adelaide, Australia, in order to gauge the degree to which aboriginal people, who have historically been marginalized based on their ethnic group and appearance in Australia, perceived discrimination to be a factor that influenced their health. Two-thirds of the participants asserted that their experiences of discrimination had influenced their health in various ways, as "some participants reported feeling physiological changes associated with their experiences of racism including increased pulse rate and blood pressure, feeling nervous, feeling nauseated, surges of adrenaline and feeling tense" (Ziersch et al., 2011, p. 1048).

Additionally, Geronimus, Hicken, Pearson, Seashols, Brown, and Cruz (2010) explored the ways in which race and health outcomes were linked. This study evaluated the degree to which Black women experience accelerated biological aging due to excessive levels of stress. According to the findings of this study, "black women are 7.5 years biologically 'older' than white women. Indicators of perceived stress and poverty account for 27% of this difference" (Geronimus et al., 2010, p. 19).

Finally, Lepore et al. (2006) assert that perceived racism experienced by Black Americans and subsequent stress responses can lead to increased levels of cardiovascular reactivity. Heightened cardiovascular reactivity has been known to lead to such forms of coronary heart disease (CHD) and cardiovascular disease (CVD) as heart attack, hypertension, and stroke.

The findings mentioned above suggest that more research needs to be done to evaluate the ways in which discrimination affects the health and well-being of underrepresented students, as well as factors that could reduce some of the physiological and psychological effects of discrimination. We share this information with you not only to impart the importance of being proactive and honest when dealing with issues of discrimination on your campus, but also to share that more research needs to be conducted for us as scholars and educators to understand how to mitigate or eradicate these health effects for students—and since someone has to do it, why not you?

UNDERREPRESENTED SCHOLARS AND TRUST

Research is a risky business. So is being an underrepresented scholar in academia. Claude Steele's work (1999) cites an important need for an emphasis on trust. As he explains in an *Atlantic* article entitled "Thin Ice: Stereotype Threat and Black College Students":

> The buildings had hardly changed in the thirty years since I'd been there. "There" was a small liberal-arts school quite near the college that I attended. In my student days I had visited it many times to see friends. This time I was there to give a speech about how racial and gender stereotypes, floating and abstract though they might seem, can affect concrete things like grades, test scores, and academic identity. My talk was received warmly, and the next morning I met with a small group of African-American students. I have done this on many campuses. But this time, perhaps cued by the familiarity of the place, I had an experience of déjà vu. The students expressed a litany of complaints that could have come straight from the mouths of the black friends I had visited there thirty years earlier: the curriculum was too white, they heard too little black music, they were ignored in class, and too often they felt slighted by faculty members

> and other students. Despite the school's recruitment efforts, they were a small minority. The core of their social life was their own group. To relieve the dysphoria, they went home a lot on weekends.
>
> I found myself giving them the same advice my father gave me when I was in college: lighten up on the politics, get the best education you can, and move on. But then I surprised myself by saying, "To do this you have to learn from people who part of yourself tells you are difficult to trust." (Steele, 1999, para. 1)

In some contexts, you may decide it makes sense to "lighten up on the politics" for your academic benefit, but, in many academic contexts, you may see the political in any action you take. Whether or not you decide to put your political and social perspectives aside in a particular situation, you can reflect on Steele's father's advice about learning and trust. Whom do you trust and whom have you trusted throughout your educational career? What happens when you learn from people you trust? How do you—and why do you—learn from people you don't trust? Keep in mind that not everyone involved in your education will have the knowledge from this chapter, that is, knowledge about the workings of different types of discrimination, stereotype threat, imposter syndrome, and solo status, as well as knowledge about available resources and strategies for addressing these issues. How might someone's level of knowledge about these issues impact how you learn from that person?

To build a trust model in undergraduate research that extends to your mentors, professors, and peers, it is important for you to know your rights and options. In a movement that might be seen as surprising to both Steele and his father, students are increasing their knowledge of politics and policies, particularly with their understanding of how politics impact the undergraduate experience. Be well versed in Title VII (Civil Rights Act of 1964), which prohibits employers from discriminating against employees on the basis of sex, race, color, religion, and national origin; Title IX (Education Amendments Act of 1972), which protects against discrimination on the basis of sex in education and educational activities receiving federal financial assistance; and the conduct codes of the universities and organizations that fund universities, including state and federal governments and private organizations. When structural protections are in place, the personal work needed for trust-building has a scaffold.

Mentoring

One of the major challenges that underrepresented students in higher education face, particularly at predominantly White institutions, is a lack of relevant academic and mentoring support from faculty members, as outlined in Chapter 4. High-achieving students from backgrounds that are underrepresented in higher education are often minorities among minorities. While students may have been celebrated or challenged by educators in their home schools and communities for their academic achievements, the scenarios to which they have been accustomed to as high school students change as they enter college. Although many professors are educated and socialized to teach their specific discipline during their graduate study, they often do not receive education that is relevant to understanding the needs of underrepresented students and helping these students overcome the obstacles that they will face. Unlike your high school teachers, most faculty members are educated to teach topics rather than to teach students. Even caring, successful faculty whose sentiments are in the right place face challenges when preparing students from specific underrepresented backgrounds for college success and graduate level achievements. Discuss your concerns with peers, faculty, administration, and alums. And if you can't find anyone to talk with, that may not be a reason to quit or leave a school or a research experience. Instead, you might make a plan for mentoring that includes multiple mentors, some within your area of study and some in a related area—or in a different area altogether—who are willing to help you navigate the research process.

The Power of the Cohort

Finding a shared experience and building trust with your peers on campus, going to see professors together, and studying and doing research together is invaluable to your research experience. Such value undergirds the concept of the Posse Foundation, a nonprofit organization that recruits and trains promising high school students from the same communities to go off to the same college together and support each other throughout the process. Your cohort doesn't have to be in your same research area or even on your same campus. Your cohort may meet you in person or virtually. We encourage you to use the power of social media to find a cohort and support for your work.

The Power of Autobiography

Sharing experiences of discrimination, stereotype threat, and solo status along with our successes helps us see our positions as changeable and transitory. Movements such as the It Gets Better campaign for LGBT youth are based on such ideas (It Gets Better Project, 2010–2016). Dr. Elaine Richardson, Professor of English and Education at The Ohio State University shared about the power of autobiography in her vignette in Charity Hudley and Mallinson (2013). She wrote:

> My educational experiences, for the most part, taught me that I was illiterate, my language was ignorant, and that I needed to get rid of my culture in order to be successful in the academy and the mainstream. There was a disconnect between what teachers knew about my home language and how this language was essential to my culture, my history, my identity, my literacy. What did those books do for people from my hood? Many of them will never open those books. Many of them will never come through these doors. I need to reach people to invite them in, validate their lives. It's taken me 14 years to get to where I am today, to write for me, my people, using my own language.

Sharing your story with others allows them to get a fuller understanding of your experiences and your perspectives on research. Encouraging others to share their full selves will give you a deeper understanding of their perspectives as well.

The Power of Reflection

As we will focus on in Chapter 7, Cress, Collier, Reitenauer, and Associates (2013) is a great guide to reflection in research where the goal of reflection is research-based action. In order to take action and even to change yourself, it is important to reflect on all that you have accomplished and the challenges you have faced. It is important to think about what you might have done differently in a particular situation and to engage your full self (emotional, intellectual, and social) to use the reflection to keep you productively moving ahead as a full person in similar future situations. It is important to see if situations you may have encountered and interpreted as individual challenges (you couldn't achieve something that you wanted) could have been a product of institutional racism (there were barriers to your success in place) in addition to individual factors (you didn't study as long as you could have). It's important to think about how such situations interact. For example, you may not have needed to study as much if you'd learned the information in a more equitable classroom. How you feel is important. Your feelings will bring about strong research questions that will lead to structural change.

The following vignette by scholar Marvin Shelton exemplifies what we mean:

> As an individual with multiple, intersecting identities, solo status has impacted my social and academic experiences in various ways. I am a Black, queer, low-income, first-generation college student, so it is not hard to imagine the extra psychological competencies that I have had to develop in order to navigate discrimination, trauma, harassment, and

isolation. I grew up in a small rural town that was homogenous in terms of population (i.e., male dominated, heterosexual, low- to middle-class, etc.) and ideology.

As a K–12 student, I faced the burden of having to navigate verbal harassment from Black and White peers around my race and my performance of race. Students believed I acted stereotypically White due to my academic and social interests. In addition, as I grew older, I began to discover that my sexual identity was not only different from those around me, but it had also seemed to place me in a subordinate position to them. Because of these negative social interactions, I hid my sexuality and tried to escape the bullying tactics my peers used to invalidate my Black identity. The defeated attitude that followed me throughout my K–12 experiences as a result of my peers' harassment was compounded by an isolating curriculum that marginalized my race, sexuality, gender-nonconforming attitudes, and, at times, my socioeconomic status. I learned, essentially, that success was White, and to be White was to be a success.

When I attended college, the marginalization and discrimination I faced as a result of my multiple identities in multiple spaces did not change. What was different, however, was the curriculum and instruction with which I was presented. I was given the opportunity to take courses that centered on my experiences through WMSURE, professors of color and queer professors, and my passion to know more about identity conflicts. As I matriculated into my university, I was introduced to Black scholars' works such as Langston Hughes's (1926) "The Negro Artist and the Racial Mountain" as well as W. E. B Dubois's *The Souls of Black Folk*. It was these works that demonstrated to me how Black scholars and artists throughout history have faced obstacles to self-definition because of dominant society's refusal to recognize the legitimacy of culturally specific paradigms. In addition, I was introduced to powerful Black, gay male characters such as Marcus in *Marcus; Or the Secret of Sweet* (McCraney, 2013) and Miss ROJ in *The Colored Museum* (Wolfe, 1988). These plays demonstrated to me that I would not be, as Marlon Riggs (Freeman & Riggs, 1989) writes in his film *Tongues Untied*, always "immersed in vanilla" when it came to the queer communities that were offered as examples in my social experiences in college. I became the center of my studies because I tasked myself with the mission to free my mind from the idea that somehow I was not beautiful, worthy, or legitimate enough to have my racial, sexual, and gendered experiences centered in the academy.

Undergraduate research offered me that opportunity to share my experiences and realize the magic of my identities through qualitative research. As a senior in college, I wrote an honors thesis where I conducted empirical research on the intersecting racial and sexual experiences of Black, gay, bisexual, transgender, and queer (GBTQ) males. This experience changed my life because I found ten other Black, queer men who were high-achieving students in the academy suffering from similar trauma produced by discrimination around their racial and sexual identities. These students shared how they felt isolated in their academic studies in terms of race and sexuality. They talked about their experiences with racism in White and gay communities as well as homophobia in Black and White communities. They also offered strategies and advice for how other Black, queer men can navigate the traditional spaces produced in the academy and society in general. I had never felt more validated in my life, as I conducted my own *mesearch*; research that included other Black, queer voices reflecting and supporting my experiences. While the literature on my topic was not extensive, I was able to build a theoretical and practical framing for my studies around the intersections of race, sexuality, gender, and social class that I have continued in my graduate studies at the University of Pennsylvania's Graduate School of Education.

Solo status is a very real factor in the experiences of marginalized students. In addition, solo status is not confined to a monolithic identity, a single space, or a particular cultural group. It is produced at the intersections of identity. It exists in classrooms and peer groups. It is a part of and produced by communities of color and queer communities. Dominant society will continue to attempt to break down marginalized identities—denying these people any freedom to self-determination—to their lowest (common) denominator.

How can we work against this limiting categorization? We must continue to strategically navigate our academic and social spaces, keeping in mind which identities we feel comfortable putting at the forefront and up for critique. This fact does not mean reenter any closets. What it means is that we all, as individuals with one or more marginalized identities, must understand that solo status and its isolating consequences are always present. Therefore, we must know that not every space will welcome all of our identities. Not every space will allow safety or comfort. Not every space will help us to develop, love, be brave, and be critical.

Therefore, we must create the safe and brave spaces. We must be the ones to carve out our spots for self-determination. We must also survive and thrive, keeping in mind that navigating discriminatory space may mean privileging one identity over the other until we can fight for and produce spaces for our full expression of self. You will find that

> time where you have learned to be fully whole. You will find those spaces to be unapologetically you. Until that time, use your writing and research as stepping stones to explore the self and theories that privilege it. Navigate your academic and social spaces strategically, always keeping your physical and mental health in the forefront. And, most importantly, know your worth and magic. Those around you may not know it yet. They are waiting for you to teach them and to read your work.

In "Thin Ice," Steele concludes, as scholar Marvin Shelton's vignette illustrates, "We cannot yet forget our essentially heroic challenge: to foster in our children a sense of hope and entitlement to mainstream American life and schooling, even when it devalues them."

Reflecting on what you are learning (content) and who you are (culture and experience) is a crucial part of the research and writing process. You can share your reflections with others or keep them to yourself. Continual reflection on the types of questions we ask throughout the chapter, such as "What is your background? What do you value? How have you made it here? Who has supported you?" allow you to reflect on your growth as an underrepresented scholar over time. We hope this chapter and your reflections on it will help you find those who believe in you and help to create others who will.

CONCLUSION

This chapter has provided an overview of the strengths that underrepresented scholars bring to undergraduate research. It is imperative for the future of higher education and for all that it affords that underrepresented scholars are more integrally represented in research in all spaces and places. Your voices matter. Thanks to all of the scholars whose lived experiences made this chapter and this book possible. We are here for you, and we cannot wait to see what you will do next.

CHAPTER 7

In Conclusion

Research in Action

> You can't pay anyone back for what has happened to you, so you try to find someone you can pay forward.
>
> —An anonymous spokesman for Alcoholics Anonymous (1944)

Research is a process rather than a finite goal. We will now reflect back on what information you brought to your reading of this book and what you learned from reading that you will now put into action. The affirmation exercise in Textbox 7.1 is a good model for how to integrate new information into research and action.

In this chapter we focus on social change, because that's what our research, from different lenses and perspectives and disciplinary approaches, focuses on. We three authors are from different backgrounds, but our interest in improving the social conditions of our students brought us together. We are fully aware that change, like the progression of research in the academy, takes generations. For, as Lily Hardy Hammond (1916) wrote, "You don't pay love back; you pay it forward" (p. 209).

The concept of "Paying it Forward" was made widely popular in the 2000 movie based on the book *Pay it Forward* by Catherine Ryan Hyde. But even before the movie, there were sayings and sentiments from leaders that encouraged us to think about education as a means to a socially just end, rather than as a goal unto itself. Nelson Mandela (2003) stated, "Education is the most powerful weapon which you can use to change the world." Likewise, John F. Kennedy (1961) encouraged us, saying, "Let us think of education as the means of developing our greatest abilities, because in each of us there is a private hope and dream which, fulfilled, can be translated into benefit for everyone and greater strength for our nation."

Many scholars have contributed through their actions. The research, teaching, mentoring, advising, and service that scholars do impact college campuses in meaningful ways. We also believe that it's okay to have multiple goals and ends. You can help people, make money, and be famous, if you so wish. But throughout it all, it's important to be reflective. It's important

Textbox 7.1: Affirmation Exercise

- What information did you already know before reading this book?
- What is the most useful thing you've learned and why?
- In what ways are you a scholar?
- How might you add to the definition of research?
- Who will work with you and support you on your journey?
- How will you disseminate information about all that you have learned and discovered?
- How can you work toward equity and justice through your research and actions?
- What information will you now put into action?
- How can you pay it forward?

to have goals for what you are doing in any particular moment—it's okay if your goals change and they truly can be yours alone. But having the goals keeps you from being caught up in other people's research arguments and academic agendas without your consent. Activity 7.1 will help you with your goals and the rationales—the sets of reasons or bases—behind your goals.

Now we turn to a different frame for looking at and reflecting on your contributions. It's the frame that we had in mind when we first decided to write this book—that is, the action of scholars who are making dedicated, innovative steps to ensure that their research has a direct impact on specific communities. We call it research in action.

From our vantage point, engagement and outreach matter in the research context because the best researchers in the world haven't figured out even some of our most immediate challenges, including poverty, homelessness, violence, and other social challenges.

Approaches to such large issues can take multiple forms. For example, an approach to ending homelessness can be historical and can motivate and inform others by showing how people have triumphed (historically and now) in different situations when faced with homelessness. Such an approach can be done through history, but also in sociology, anthropology, psychology, theatre, music, art, and an interdisciplinary mix of them all!

In social movements around the world, particularly in the 2 years leading up to this book's publication, students have been sharing their experiences and speaking out against injustice on campuses and in the greater world. Such work is frequently done through mentoring others. We want to especially encourage you to reach out to high school and younger students to set them on a path toward being scholars in college as well.

Activity 7.1: What's Important to You? What Are Your Goals?

One thing that you did when you applied to your undergraduate institution was describe yourself, your experiences, and your goals. This writing experience likely helped you reflect on what was important to you. Graduate schools will also require that you write a statement in which you describe what is important to you and what your goals are. This statement allows the faculty in the program to get to know your strengths and what you are looking for in a graduate program and a career. Demonstrating that you have strong, attainable goals and that you have accomplished some of your past goals will help the admissions committee see you for the scholar that you are. Reflecting on the questions below throughout your college experience may give you some direction and allow for periods of reflection about the past and planning for the future. Thinking about the answers to these questions can also help you prepare for those graduate applications in a few years.

- What are your goals as you read this book?
- Why have you chosen to make them your goals?
- Why do you study what you study? What do you enjoy about what you study?
- Why do you want to be good at what you study?
- What are you currently good at? In which areas do you excel?
- What might your goals be 5 years from now?
- How have your goals changed over the last few months? Years?
- Are there people or communities that you would like to help?
- How would you most like to give back to your own community?

In addition to promoting change among those around you, you can also make changes at a broader level. Even as undergraduate scholars, you can change the research paradigm as you go along. To help you with this challenge, we now use examples from Banks (2016) to illustrate how you as an undergraduate scholar may engender social change through research and education. Too often, higher education is dominated by a "contributions" and/or an "additive" approach to diversity (Banks, 2016), whereby cultural content is limited primarily to special ethnic events and celebrations (e.g., Black History Month), and cultural content is added without restructuring the curriculum. In what Banks (2016) calls "transformative" and "social action" approaches, however, the basic assumptions and the structure of the curriculum are changed: Students are taught to view concepts, issues, events, and themes from the perspectives of diverse social and cultural groups and to become reflective social critics and skilled participants in social change. Such approaches are alive with the Black Lives Matter movement as well as with the demands that students are making to change college campuses, found online at www.thedemands.org.

The Social Action Approach in the multicultural education movement exemplifies why research in action is so crucially important. In both a research and a Social Action Approach model, scholars make decisions on important social issues and take research-based actions to address them. The Social Action Approach further requires students to make decisions and take actions related to the concepts, issues, or problems studied—approaches that are important to the research process (Banks, 2016).

The Social Action Approach also empowers you and helps you acquire political efficacy. Too often the traditional goal of school has been to socialize students to assimilate into the dominant school culture so they will accept unquestioningly the existing ideologies, institutions, and practices of a given society. But now there are increasingly more courses and experiences that can help you to become researchers for change. You can even have a hand in designing and teaching these courses.

Service learning is a way to make social action a crucial component of your education. The Corporation for National and Community Service (2008) defines service learning as a method of teaching "curriculum-based community service that integrates classroom instruction with community service activities" (p. 13). Service learning as an extension of collegiate volunteerism is an upward trend. At least a quarter of all higher education institutions and more than half of all community colleges have adopted service-learning programs (Corporation for National and Community Service, 2008). Talk to faculty or the advising office at your school to learn if your college offers service-learning courses. To learn more about service learning, whether or not you take such a course, we highly recommend the book *Learning through Serving: A Student Guidebook for Service-Learning and Civic Engagement across Academic Disciplines and Cultural Communities* (Cress et al., 2013), which introduces readers to elements of service learning such as building community partnerships, reflection, leadership, failure, and how to evaluate your work beyond a grade.

Action research is an approach that focuses on action and research simultaneously and in a participative manner (Coghlan & Brannick, 2010). Just as in community-based participatory research (below), there are several approaches to action research, but all have this in common: Research participants are themselves researchers or involved in a democratic partnership or co-researcher model with a researcher. Research is seen as an agent of a specific stated change. As such, data and other forms of findings are generated with input of participants who become integral research participants.

Similar to action research, community-based participatory research (CBPR) is an applied collaborative approach that enables members of a given community to actively participate in the complete process of research. Community members are involved in the research—from conception of the ideas, to the research methodology and design, to execution of a research plan, to the analysis of data, to the interpretation of data, to the development of conclusions, to the dissemination and communication of results—with a goal of influencing positive change to meet specific community needs (National Institute on Minority Health and Health Disparities, n.d.).

In the undergraduate context, CBPR expands the service-learning teaching approach to focus in on research in addition to direct service and action. The CBPR approach is beneficial research because all participants are included so that researchers are working with rather than for or on communities. For example, Prof. Charity Hudley's former honors thesis student Ms. Rachel Brooks researched multicultural education in high school English contexts; she interviewed teachers to include their perspectives on how to incorporate multicultural education into policy.

Scholar Amirio Freeman explains how CBPR is important to his research:

> One of the best things about spearheading your own research project is that you have the opportunity, through your project, to make the world a more just and equal place. For instance, a research effort may have the potential to create a more equitable global society by spotlighting an overlooked social problem or even by offering a solution to a long-standing societal ill. However, even though research has the ability to mitigate inequality, research itself may *be* a source of inequality. In many cases, research projects—especially those that involve working with a specific community of individuals—generate unevenness: a researcher may derive more benefits from a research project than the community that contributed to the project; some populations may have more access to a set of research results than others; and so on. As a student who is interested in the aforementioned ability of research to create social change, doing research in a way that lessens power and privilege differentials is always a top priority for me, which is why I utilize the community-based participatory research method with my own work.
>
> Right now, I am in the midst of completing an honors thesis project that will involve collecting oral histories from individuals in my hometown. Through using oral histories to complete my project, I am placing myself and my community—the central subject of my research—on equal footing in many ways, helping to create a true sense of collaboration between myself and those that I am working with. For example,

the oral histories I will conduct will not only provide material for my honors thesis, but the oral histories will also be useful to my community since I will be archiving and collecting narratives that may otherwise be forgotten. This, as a result, creates evenness when it comes to who will benefit from my work. Overall, this use of the CBPR method is so exciting for me because the method helps to ensure that I am performing research in a way that aligns with my belief in always navigating the world in a just, fair way.

DISSEMINATION

The most important aspect in the broadening of the academy is the dissemination of information to those who have not had access to it in the past. For example, sharing information about how to get to college and how to go succeed in it, how to do research, and how to then share that research information after it has been created is an important aspect of this dissemination process. So it is important to continually ask yourself who has access to the knowledge in research papers and articles. What's the process by which educators, students, parents, and others in a particular community can access information that is created and published by researchers?

We asked this question of ourselves as we thought about who could even access this book. It was really important for us, for example, to ensure that the book had a low sale point (cost to you) and would be published by a publisher with a broad distribution base that would publish the book in ebook form as well as in paperback. To make that process possible, we had to ourselves pay what is known as a subvention, money that each of us paid out of our own pockets ($5,000 total) to ensure that there would be a lower publishing price than the one naturally set by market forces. In addition, we do many talks at high schools and universities, educational organizations, and academic societies in order to disseminate what we've learned and encourage students to become scholars. We encourage you to do the same!

Important Aspects of Dissemination. Dissemination of information outside of scholarly environments does not happen as often as it should in academia. Academic research is usually valued over writing that is designed for more general readerships. In addition, the skills needed to write for a broad audience might not be addressed in PhD programs. Although interesting and important research findings are written up and published in academic forums like journal articles and books all the time, dissemination of this work to a more general audience often is not a priority. It is well documented that sharing the research that occurs in the academy with the public as well as with policymakers, businesses, and practitioners is not a given. There are several reasons for this. First, many academics write in their own language varieties—recall our discussion of jargon in Chapter 5. Academic writing can sometimes be understood only by a small group of people; others may not be able to translate the research they read, no matter how interesting or relevant they may find the work. Along these same lines, researchers may worry that the media or the public will misinterpret their findings. When reporting the results of a study, for example, a media representative may inaccurately summarize the findings because of the use of jargon or their lack of expertise in that area.

Having research misrepresented by others or the media can have important negative consequences for academic researchers. In addition, researchers often caution about making too much of their findings public. When they publish a research study, for example, it is common for researchers to say that more research is needed before conclusions can be made. For example, a single paper reporting that a certain treatment can improve outcomes in one group of patients does not provide enough evidence to conclude that the treatment should be used for everyone with that particular malady. Researchers are therefore hesitant to share their findings with the public when in fact more research needs to be done to draw stronger conclusions. Finally, there is not much incentive for academics to share their findings with the public. The premium in the academy, and what typically earns someone tenure and respect in their field, is still focused on publishing highly specialized and often technical articles for a scholarly audience. So there is a real competition for time and energy. Many of us make our reputation and our living by sharing our work with the small group of experts in our field in the publications that these experts will read and at the conferences they will attend. There are rarely incentives for sharing research findings with the public, however useful this may be. Dissemination can be a particularly tricky issue because it requires its own brand of scholarship to figure out how to reach specific audiences, motivate communities, and impact policy. And the system that's already in place tends to win. But times are changing. So we present here several approaches to changing this situation in the academy. We're doing our part to reframe the system, and we hope you will too.

In contrast to traditional academic scholars, *public scholars* are researchers dedicated to making scholarship accessible by writing and developing materials for a broad audience. Public scholars both create their own research and findings and summarize and synthesize the work of others so that it is more easily read and understood by others. Public scholars are often supported by professional organizations as well as by federal agencies such as the National Endowment for the Humanities (NEH). In writing this book, we are in a sense public scholars because we're writing directly to you rather than for your professors or for a scholarly audience. Social media has greatly enabled the impact of public scholars.

Public intellectuals are those who bring research findings into the public discourse and comment widely on a spectrum of issues and ideas. Some are researchers in their own right, and all are thinkers and synthesizers. Many start out in a specific topic area and then branch out to comment and think about a broader range of topics. Public intellectuals share an academic lens on broad concepts. But as theory purveyance grows, there's the danger of a loss of dedicated research experience on particular topics. The opinions of public intellectuals are important, however, as no one has time to research everything, yet we have to think and process information on a wide range of topics just to stay informed about what is going on in the world.

The National Science Foundation, a major source of funding for natural and social scientists in particular, has responded to the need for more direct commitment to public interest and has called for more initiatives to add a direct service and public scholarship component to scholarly research. NSF proposals must include an explanation of the "broader impacts" of the proposed activity. Under this initiative, the NSF requires answers to the following questions:

- What are the broader impacts of the proposed activity?
- How well does the activity advance discovery and understanding while promoting teaching, training, and learning?
- How well does the proposed activity broaden the participation of underrepresented groups (e.g., gender, ethnicity, disability, geographic, etc.)?
- To what extent will it enhance the infrastructure for research and education, such as facilities, instrumentation, networks, and partnerships?
- Will the results be disseminated broadly to enhance scientific and technological understanding?
- What may be the benefits of the proposed activity to society? (www.nsf.gov/pubs/gpg/broaderimpacts.pdf)

Even though gathering new knowledge is deemed necessary for a research endeavor, the NSF asks for plans for immediate dissemination of the knowledge. These questions also promote awareness on the part of the researcher of the more immediate value of dissemination versus adhering to a more typical research and dissemination process, which would expect dissemination to happen some time after research information has been acquired. In addition, the NSF now values products (videos, podcasts, images, and other media) in part so that there are more accessible findings from scholarly research.

CHALLENGES TO COMMUNITY RESEARCH

A challenge to the work that we are encouraging you to do throughout this chapter is the way that action and community research has often been devalued in academia or seen as extra or bonus work that is nice when it happens but is not essential. This viewpoint devalues community input and keeps academia isolated. Yet such work is more popular among underrepresented scholars, so we want to caution you to beware of the marginalization and the silent devaluation of the research approaches of those who are underrepresented in academia. Devaluation of community-based research is academic discrimination. We urge you not to replicate those discriminatory actions without question. For example, someone might say, "I think the research should be conducted this way because that's how it's always been done methodologically." Other phrases you may hear might include: "You should frame the question this way," "These are the topics that we think about in this discipline," "This is not linguistics/psychology/science/etc.," and "This is great but this is not research." We contend that if we try to work comprehensively to open up the academy, we must include not only those who don't have a similar perspective, skin color, or home country, but also those who truly have different academic values and beliefs. If you don't have answers for why, why replicate discriminatory actions? Paris and Winn (2012) encourage scholars from underrepresented backgrounds to tangle with such questions—to understand that the work is strong, immediate, forceful, and that it includes feelings. The narration of the research process is so important. They encourage us to humanize research. For example, the creation of a community studies minor and Prof. Charity Hudley's accompanying community studies professorship was the backbone of the support for WMSURE and this book.

RESPONSIBILITY MATTERS IN RESEARCH

The principle of *debt incurred* (Labov, 1982) contends that intellectually you *should* be paying information forward—through research and writing as *responsibility*, which is part of the framework of social justice research.

Responsibility is based on this history and trajectory of what's happening across communities. All kinds of research are needed to address community needs and interests. In Activity 7.2, we want you to think about what your responsibilities are.

We encourage you to think locally first, and of your own college campus. A localized model encourages you to think about the responsibility that you have to those right around you. You can start with learning about research as a responsibility in Activity 7.3.

Activity 7.2: Thinking About Your Responsibilities

- What are your responsibilities?
- To yourself?
- To your families?
- To your communities?
- To those you live with?
- To those you do research with?
- To those who taught you?

Importance of Spreading the Word and How to Do So Through Peer Mentorship

These questions about sharing information about research are immediate and far reaching, as they ask us to critically articulate whom we want to benefit directly from our work. One way to start now is to be a peer mentor. A peer mentor is a student who can serve as a resource, a helping hand, a soundboard, and a referral service. Peer mentors provide support, encouragement, and information to students. Because peer mentors are students and are typically senior to you, they have been in your position and thus have experience that is directly related to what you will encounter. You can learn from their mistakes and from their successes, and take the advice that they feel you will benefit from. Mentors are typically academically successful and serve as role models for what mentees can achieve. In the best cases, mentors can provide a connection between a mentee and the larger community.

There may be a formal peer-mentoring system at your school through which you can be assigned a mentor. Regardless of whether you have an official mentor, you may want to seek out an informal mentor. You do not necessarily have to be struggling to benefit from working with a peer mentor. In our experience, some of the most successful students are those who have benefited from positive interactions with upperclass students who have excelled at research or have earned honors. Establishing a relationship with a peer mentor who can give you tips for success can make the difference between being a student and being a scholar. Peer mentors have been in your shoes recently, whereas faculty mentors most likely went to college at least 5–6 years earlier. Potential mentors are all around you. If you are starting to conduct research and are working in an environment with a research group, an obvious candidate for a peer mentor would be someone who has been working in this group for a while. Good peer mentors will likely have expertise not only in the research you are working on but also on the broader

Activity 7.3: Learning About Research as a Responsibility

- What is taught to incoming students about research?
- What is taught to faculty about undergraduate research and opportunities both on and off campus?
- What is taught to everyone else, including parents, about the value of research? Are secondary school students and teachers getting educated about research?
- What examples of research are shared with others?

research topic. They will most likely share the same major and therefore will have taken some of the same classes that you will need to take.

Students who have worked in a research setting have learned how to balance classes, research, and social life and thus serve as excellent mentors. Feel free to approach one of these students and ask them specific as well as general questions about their experiences. Most of these peer-mentor relationships are informal and can develop just by your seeking out this person for advice and checking in from time to time. You may be surprised to learn how eager individuals are to share the keys to their success! If you don't feel comfortable seeking out your own mentor, you can ask your faculty advisor to recommend someone from the group who has been particularly successful.

Importance of Serving as a Peer Mentor

Once you establish yourself as a scholar at your institution, you can act as a mentor to other students. Mentors benefit from interactions with their mentees in that they typically develop positive friendships with them and derive satisfaction from helping someone and changing their life for the better. Peer mentors *see, hear, and understand things that faculty and staff cannot.*

If you are interested in mentoring younger students and there are no formal mentoring programs at your college, you can seek out students on your own, or even look into developing a mentoring program yourself, say within your major or your research lab. The students you seek out may be individuals who are part of your social group or someone you notice is struggling in a particular setting. Reaching out to someone in these situations can establish the beginning of a relationship that may lead to something more long-term. The students in our program who have navigated through their classes, research experiences, and college life often have informal mentoring relationships with students who are taking similar classes or who are just starting research. They are some of the students who have provided their thoughts in vignette form throughout this book. They re-

Activity 7.4: Whom Can You Help/Educate/Teach/Mentor Now That You Have Read This Book?

- What audience of your peers would you most like to reach?
- What life or academic skills do you have that you could pass on to other students?
- What lessons have you learned regarding academics or your personal life that you wish you had learned earlier?
- What parts of this book gave you the most insight?

port to us that mentoring can have many benefits and is a valuable part of their academic and personal development. You can take parts of what you learned from this book to help you mentor others (try Activity 7.4). There may also be paid opportunities for you to mentor students by working as a peer tutor or through another program on campus. You can also ask your classmates even without a formalized structure, "How can I help you succeed?" Cress et al. (2013), referred to above, has a chapter on mentoring with frameworks that can help you make the most out of a peer-mentoring relationship.

Importance of the Professorate

College is the gateway to the professional life, as we described in Chapter 1. Higher education needs people from all backgrounds involved in that gateway process. Faculty have different reasons for why they choose to be professors. Some become professors because they have a great love for their academic specialty, others because they have a great love for students, and many become professors because they love a life of learning. None of these perspectives is exclusive, and your professors may feel motivated by different forces at different times in their careers. What is important for you is to start having conversations with faculty about how their reasons for being professors intersect. As you learned in Chapter 4, faculty at research universities have a focus on both research and teaching. Liberal arts faculty may focus on teaching and may also focus on research. Community-college faculty may have a focus on teaching and also do research. Some faculty teach as adjuncts. Some adjuncts may work at multiple schools even in one semester. Other adjuncts may work other jobs and may teach at night or online. Different types of schools have different missions and different job requirements, which broadens your opportunities to find something that best suits you. Activity 7.5 is designed to get you talking to professors about why they do what they do.

Activity 7.5: Why Professors Do What They Do

Meet with your favorite professor(s) and ask them the following questions:

- Why did you become a professor?
- What kept you happy being a professor for 5 years? 10? 15?
- What is the best part of being a professor?
- What do you not like about being a professor?

No matter your ultimate personal and professional goals, research gets you on the pathway to graduate and professional school. And that's why obtaining a higher degree is important—so that knowledge as well as practice and policy are diverse—and you get to do exciting teaching and research!

The salary ranges for public university professors are easily found online, and the *Chronicle of Higher Education* provides salary ranges for professors at specific colleges, college types, and states; look up these data online at data.chronicle.com. Keep in mind that many PhD programs offer tuition and other support for highly qualified candidates and that faculty can earn additional money from research grants and by teaching in the summer. As an added bonus, many colleges and universities provide opportunities for travel to conferences, to present research at other institutions, and for professional training. In fact, colleges provide funds and want professors to travel, experience new things, and take students with them—learning with students keeps us current! For most faculty, money is just part of the equation; many of us are very happy because we get to work with students and do what we love.

Yet there are many people who may question your interest in becoming a professor. So here we give you some support on how to deal with people who don't understand your interest in the professorate—including family, friends, etc. Instead of disagreeing with them, we suggest that you find out what they value and how education helped to make their journey possible. We also suggest that you show what is great and fun and what is tough about being a professor so that they know that you have made an informed choice. Our answers will also help you respond to parents and guardians, friends, and even professors who dissuade you from entering academia. Below are some themes to focus on and some narratives that illustrate these themes.

- Why being a professor is so important for you and for the world.
- Why we need academically high-achieving professors in the classroom, in leadership, and in research.

- How much money professors actually make and what they can do to increase their income.
- Why money isn't everything and why those who do work that we do in places that we do walk around so happy.

For example, below we list some fun stuff:

- Professors often get to set their own hours. So for Prof. Charity Hudley, that means she teaches in the afternoon and writes in the mornings. For Prof. Dickter, this ensures she can go to the gym every morning before heading to work. For many faculty, it means that they can be there in the morning to put their kids on the bus and they have the flexibility to go to school or sporting events when they come up.
- Professors get to study what they want! There is a large range of topics. You can really follow things that you're interested in. You have a chance to decide what you're most interested in, and academia will give you that choice even from the beginning of your career. Prof. Dickter didn't start doing research on the effects of parental smoking until she was several years into her tenure-track job. Since her father smoked his entire life, she had always been interested in the cognitive mechanisms that led children of smokers to be more likely to smoke themselves, but it wasn't until she began talking to a colleague about this interest that it developed into a research program and a formal lab that is funded by the National Institutes of Health.
- After about 7 years, tenure-track faculty earn tenure and their jobs are secure. The tenure process may vary by school but faculty are usually supported by peers and mentors and afterward they have greater freedom to say what they want and potentially take more risk for that part of their career. Other faculty are given 3- to 5-year contracts for stability.
- Faculty get to travel and meet lots of new people! Faculty travel to conferences, to research sites, and to other places in the community including schools, and, for some, this is definitely one of the best parts of the job.
- Faculty get to work with students and are constantly learning and sharing what they learn! Teaching and research keep faculty active; there are always new ideas, new methods, new themes, new projects, and great new ideas from students. You use your mind in so many ways that the job never becomes boring! Prof. Dickter enjoys that students each bring their unique perspectives with them to the research lab, which keeps her research, as well as the research in the field, fresh, and opens more doors into new areas of research for her.

Now we list the reality of some tough stuff:

- Faculty often have to work long hours, although those hours may be flexible in terms of what time of the day you're working. In the early years when you're planning courses and research from the start, the hours are longer. Even though faculty have flexibility in when they work, there is a lot of work.
- Faculty have to write on a regular schedule in order to publish. For many faculty, especially when they are just starting out, the load of writing is the hardest part. We suggest thinking about yourself not just as an educator or a teacher, but as a writer too, and figuring out a regular schedule that works for you, whether it's daily, weekly, monthly, or yearly. (Prof. Charity Hudley prefers to do the bulk of her writing in the coldest of the winter and hottest of the summer months.)
- Faculty pay is lower than that of professionals in the corporate sector with the same level of education. This is definitely true early on, but the flexibility and benefits may make up some or all of the difference.
- It is often hard to get a job in the location that faculty want. We suggest establishing relationships in areas that you're interested in: email professors, read up on their work, give talks, or visit, and so forth. If you can't visit places that you'd like to attend for graduate school before you apply or attend, you can work to get a sense of the types of universities that are available in the places that you would like to live so that you can both tailor your interests to particular university needs and encourage them to need what you have to offer.
- Another hard part is figuring out how to play the game of the academy so that you don't have the imposter syndrome we refer to in Chapter 6. In other words, there are lots of strategies to help you become a professor and earn tenure that you may not necessarily learn in graduate school. That's why we wrote this book—to share the strategies you can use beginning with your undergraduate experience! There are more and more resources for aspiring and junior faculty so that this "hidden curriculum" is becoming less hidden.

Through research we want you to get the skills you need to be scholars and leaders early on so that you have longer to share them with others!

We also want to implore you: Don't believe the hype that it's not worth it for underrepresented scholars to become teachers and professors—it's especially important for underrepresented scholars to become teachers and professors for reasons of social and educational justice, and the numbers are growing very slowly. Research from major educational organizations including the Albert Shanker Institute (Watson, Bristol, White, & Vilson,

2015) and the Stanford Center for Opportunity Policy in Education (Bristol & White, 2015) shows that recruitment of educators from underrepresented backgrounds has made some progress, but retention still lags. In K–12 education, 43 percent of students and 17 percent of teachers are people of color (Hixenbaugh, 2011). Higher education remains largely racially segregated. For example, at the College of William & Mary, while the student body is made up of 33% students of color, 511 of the college's 632 full-time instructional faculty are White (Boyle, 2014).

Without more educators whose backgrounds and identities are fairly represented in a socially just educational system, we'll never make the changes needed. And the transformations of lives that we three have seen won't be spread and hope will be lost. The U.S. educational system, and its higher-education system in particular, is still the nexus for economic and social achievement. We all need to be active participants in that process.

Research has also shown that an increase in educators from underrepresented backgrounds enriches the learning experience of students from all backgrounds, as the importance of lived experience serves to increase students' ability to empathize and learn from those who are not like themselves.

It is important to think about how to respond to people who may dissuade you from education or academia. We encourage you to share the information in this chapter with them and have them think of the teachers that helped them get to where they are and that helped others who needed teachers even more. We encourage you to ask others—if someone has to do this work for justice, then who should it be?

TOPICS FOR THE FUTURE OF UNDERGRADUATE RESEARCH

In conclusion, we are interested in what you think we should address in the revision or second edition of this book. For us, the research and development of this book continues. We're working now to expand our model of undergraduate research across William & Mary through a college-wide enhancement of teaching at William & Mary and across the Commonwealth of Virginia through events that bring students and faculty together to talk about undergraduate research and to learn from each other.

In the future we plan to broaden the conversation to even more states and universities in structural ways. Our plan includes working with faculty and administrators to educate them about undergraduate research so that they are better prepared for the experience. Our work will also help them rethink how undergraduate research is structured on campus—both in classes and outside of classes—and how students earn funding and credit and how faculty earn funding and merit.

To make such changes, we need your input so that we can engage in a community-based participatory research model, as we have with our students in writing this book.

You will have this information earlier than others and can help to make it more a part of the academy—undergraduate research as a value. Your participation in organizations including the McNair Scholars Program and the Council on Undergraduate Research will provide models for others.

We also want your input on the best practices for growing an undergraduate research program or for starting one. Send us models of undergraduate research programs and keep us updated on the state of the art. Most of the information in this book came from an undergraduate research program we created (WMSURE), and we are always looking for ways to improve and grow our own program as well as keeping the national and international conversation going.

We encourage you to see our website and to contact us on social media for more information, including additional vignettes, photographs, videos, and interactive surveys.

Join us online to share your experiences with others:

- Facebook: www.facebook.com/undergraduateresearchguide, in addition to: www.facebook.com/Dr.AnneH.CharityHudley
- Twitter: @underesearch, in addition to @acharityhudley, @cheryldickter, and @hannahafranz
- Websites: undergraduateresearchguide.com/ and charityhudleydickter-franz.com
- Instagram: instagram.com/undergraduateresearchguide
- Google+: undergraduateresearchguide@gmail.com

In this book, we have exemplified how you can be a part of this work that will indeed take generations. We cannot wait to see all that you come up with and to read, listen to, see, and feel the impact of your research both on you and on those around you.

References

Abela, C., & Renfro, T. (2001). *Time management for college students* [Powerpoint slides]. Retrieved from www.roanoke.edu/Documents/clt/Time%20Management%20presentation.pptx

Abela, C., & Renfro, T. (2002). *Top ten misconceptions students have about college*. Retrieved from gorpcc.com/library/wp-content/uploads/2013/06/Top-ten-misconceptions-about-college-TRIO.pdf

Abrams, P. (Producer), & Leder, M. (Director). (2000). *Pay it forward* [Motion picture]. United States: Warner Bros.

Akinsola, M. K., Tella, A., & Tella, A. (2007). Correlates of academic procrastination and mathematics achievement of university undergraduate students. *Eurasia Journal of Mathematics, Science & Technology Education*, *3*(4), 363–370.

American Economic Association. (2016). *American Economic Association*. Retrieved from www.aeaweb.org

Andrew W. Mellon Foundation. (2016). *Diversity*. Retrieved from mellon.org/programs/diversity/

Andrew W. Mellon Foundation. (2016). *Mellon Mays Undergraduate Fellowship Program*. Retrieved from mellon.org/programs/diversity/mellon-mays-undergraduate-fellowship-program/

Association for Psychological Science. (n.d.). *Association for Psychological Science*. Retrieved from www.psychologicalscience.org/

Association of American Colleges & Universities. (n.d.). *Making excellence inclusive*. Retrieved from www.aacu.org/making-excellence-inclusive

Aud, S., Fox, M. A., & KewalRamani, A. (2010, July). *Status and trends in the education of racial and ethnic groups*. Washington, DC: National Center for Education Statistics Institute of Education Sciences.

Australian Journal of Career Development. (2011, April). (20)1 [Entire issue]. Retrieved from acd.sagepub.com/content/20/1/32.short

Ayers, R., & Ayers, W. (2011). *Teaching the taboo: Courage and imagination in the classroom*. New York, NY: Teachers College Press.

Bain, K. (2012). *What the best college students do*. Cambridge, MA: Belknap Press of Harvard University Press.

Banks, J. A. (2016). Approaches to multicultural curriculum reform. In J. A. Banks

& C. M. Banks (Eds.), *Multicultural education: Issues and perspectives* (9th ed.) (pp. 151–170). Hoboken, NJ: John Wiley & Sons.

Basken, P. (2012, February 13). National Science Foundation steps up its push for interdisciplinary research. *Chronicle of Higher Education.* Retrieved from chronicle.com.proxy.wm.edu/article/National-Science-Foundation/130757

Bauer, K. W., & Bennett, J. S. (2003). Alumni perception used to assess undergraduate research experience. *Journal of Higher Education, 74*(2), 210–230.

Belcher, W. L. (2009). *Writing your journal article in twelve weeks: A guide to academic publishing success.* Thousand Oaks, CA: SAGE Publications.

BlackLivesMatter. (n.d.). *Black lives matter.* Retrieved from blacklivesmatter.com/

Boone, L., Soenens, B., Braet, C., & Goossens, L. (2010). An empirical typology of perfectionism in early-to-mid adolescents and its relation with eating disorder symptoms. *Behaviour Research and Therapy, 48*(7), 686–691.

Booth, W. C., Colomb, G. G., & Williams, J. M. (2008). *The craft of research.* (3rd ed.). Chicago, IL: University of Chicago Press.

Boyle, A. (2014, November 18). At least 80 percent of the college's faculty are White. The Flat Hat. Retrieved from flathatnews.com/2014/11/18/at-least-80-of-the-college-faculty-are-white/

Bristol, T., & White, T. (2015, September 10). What can districts do to recruit more teachers of color? *Stanford Center for Opportunity Policy in Education* [Blog]. Retrieved from edpolicy.stanford.edu/blog/entry/1373

Buehler, R., Griffin, D., & Ross, M. (1994). Exploring the "planning fallacy": Why people underestimate their task completion times. *Journal of Personality and Social Psychology, 67*(3), 366–381.

Busteed, B. (2015, April 18). Is college worth it? That depends. Gallup. Retrieved from www.gallup.com/opinion/gallup/182312/college-worth-depends.aspx

Cabrera, A. F., Crissman, J. L., Bernal, E. M., Nora, A., Terenzini, P. T., & Pascarella, E. T. (2002). Collaborative learning: Its impact on college students' development and diversity. *Journal of College Student Development, 43*(1), 20–34.

Carter, J. J. (2012). Multicultural science education. *College of William & Mary Undergraduate Honors Theses.* Paper 554. Retrieved from publish.wm.edu/honorstheses/554/?utm_source=publish.wm.edu%2Fhonorstheses%2F554&utm_medium=PDF&utm_campaign=PDFCoverPages

Charity Hudley, A. H. (2008). Linguists as agents for social change. *Language and Linguistics Compass Sociolinguistics Section, 2*(5), 923–939.

Charity Hudley, A. H. (2012). Linguistics and social activism. In R. Bayley, R. Cameron, & C. Lucas (Eds.), *The Oxford handbook of linguistics.* Oxford, UK: Oxford University Press.

Charity Hudley, A. H., Harris, J., Hayes, J., Ikeler, K., & Squires, A. (2008). Service learning as an introduction to sociolinguistics and linguistic equality. *American Speech, 83*(2), 237–251.

Charity Hudley, A. H., & Mallinson, C. (2013). *We do language: English language variation in the secondary English classroom.* New York, NY: Teachers College Press.

Charity Hudley, A. H., & Mallinson, C. (2010). *Understanding English language variation in U.S. schools.* New York, NY: Teachers College Press.

Charity Hudley, A. H., & Mallinson, C. (Eds.). (in press). Linguistics and the broader university. Special Issue of the *Journal of English Linguistics.*

Civil Rights Act of 1964 § 7, 42 U.S.C. § 2000e et seq. (1964).

Claessens, B. J. C., van Eerde, W., Rutte, C. G., & Roe, R. A. (2007). A review of the time management literature. *Personnel Review, 36*(2), 255–276.

Coghlan, D., & Brannick, T. (2010). *Doing action research in your own organization* (3rd ed.). London, United Kingdom: Sage.

Cohen, G. L., Garcia, J., Apfel, N., & Master, A. (2006, September 1). Reducing the racial achievement gap: A social-psychological intervention. *Science, 313*(5791), 1307–1310.

Cokley, K. (2006). The impact of racialized schools and racist (mis)education on African American students' academic identity. In M. G. Constantine & D. W. Sue (Eds.), *Addressing racism: Facilitating cultural competence in mental health and educational settings* (pp. 127–143). Hoboken, NJ: John Wiley & Sons.

College Board. (2016). *AP Research: Course and exam description* (2nd ed.). New York, NY: CollegeBoard. Retrieved from secure-media.collegeboard.org/digitalServices/pdf/ap/ap-research-course-and-exam-description.pdf

Corporation for National and Community Service. (2008). *Community service and service-learning in America's schools.* Washington, DC: Corporation for National and Community Service.

Cosgrove, T. J. (1986). The effects of participation in a mentoring-transcript program on freshmen. *Journal of College Student Personnel. 27,* 119–124.

Council on Undergraduate Research. (2016). Council on Undergraduate Research: Learning through research. Retrieved from www.cur.org/

Crocker, J., & McGraw, K. M. (1984). What's good for the goose is not good for the gander: Solo status as an obstacle to occupational achievement for males and females. *American Behavioral Scientist, 27*(3), 357–370.

Cress, C. M., Collier, P. J., Reitenauer, V. L., & Associates. (2013). *Learning through serving: A student guidebook for service-learning and civic engagement across academic disciplines and cultural communities* (2nd ed.). Sterling, VA: Stylus Publishing, LLC.

Dahlvig, J. (2010). Mentoring of African American students at a predominantly White institution (PWI). *Christian Higher Education, 9,* 369–395.

Dartmouth College Academic Skills Center. (2001a). *How well do you plan?* Retrieved from www.dartmouth.edu/~acskills/success/time.html

Dartmouth College Academic Skills Center. (2001b). *The master schedule.* Retrieved from www.dartmouth.edu/~acskills/success/time.html

Dasgupta, N. (2011). Ingroup experts and peers as social vaccines who inoculate the self-concept: The Stereotype Inoculation Model. *Psychological Inquiry, 22*(4), 231–246.

Dickter, C. L., Charity Hudley, A. H., Franz, H., & Lambert, E. (in preparation). Assessing and mitigating solo status and stereotype threat among high achieving underrepresented students: A mixed-methods approach.

Dodge, L., & Derwin, E. B. (2008). Overcoming barriers of tradition through an effective new graduate admission policy. *The Journal of Continuing Education, 56*(2), 2–11.

Dosch, M., & Zidon, M. (2014). "The course fit us": Differentiated instruction in the college classroom. *International Journal of Teaching and Learning in Higher Education, 26*(3), 343–357.

Dubois, W. E. B. (1903). *The Souls of Black folk: Essays and sketches*. Chicago, IL: A. C. McClurg & Co.

Dubois, W. E. B. (1903). "The talented tenth." In *The Negro problem: A series of articles by representative American Negroes of to-day*. New York, NY: James Pott & Co.

Economic Science Association. (2014). *Economic Science Association*. Retrieved from www.economicscience.org/index.html

Elbow, P. (2012). *Vernacular eloquence: What speech can bring to writing*. New York, NY: Oxford University Press.

Eliasson, A. H., Lettieri, C. J., & Eliasson, A. H. (2010). Early to bed, early to rise! Sleep habits and academic performance in college students. *Sleep and Breathing, 14*, 71–75.

Ellison, J., & Eatman, T. K. (2008). *Scholarship in public: Knowledge creation and tenure policy in the engaged university: A resource on promotion and tenure in the arts, humanities, and design*. Syracuse, NY: Imagining America: Artists and Scholars in Public Life. Retrieved from imaginingamerica.org/wp-content/uploads/2015/07/ScholarshipinPublicKnowledge.pdf

Fabbri, M., Mencarelli, C., Adan, A., & Natale, V. (2013). Time-of-day and circadian typology on memory retrieval. *Biological Rhythm Research, 44*(1), 125–142.

Francis-Smythe, J. A., & Robertson, I. T. (1999). On the relationship between time management and time estimation. *British Journal of Psychology, 90*, 333–347.

Franz, H. (2016, December). *Access to college-ready writing through classroom writing assessment*. Paper presented at the annual conference of the Literacy Research Association, Nashville, TN.

Freeman, B. (Producer), & Riggs, M. (Director). (1989). *Tongues untied* [Motion picture]. United States: Signifyin' Works.

Freire, P. (1970). *Pedagogy of the oppressed*. New York, NY: Herder and Herder.

Geronimus, A. T., Hicken, M. T., Pearson, J. A., Seashols, S. J., Brown, K. L., & Cruz, T. D. (2010). Do US Black women experience stress-related accelerated biological aging? A novel theory and first population-based test of Black-White differences in telomere length. *Human Nature, 21*, 19–38.

Giovanni, N. (2007). *Acolytes*. New York, NY: HarperCollins.

Gray, D. E. (2014). *Doing research in the real world* (2nd ed.). Thousand Oaks, CA: Sage.

Group to combat alcoholism grows apace in anonymity. (1944, January 8). *Christian Science Monitor*, p. 3.

Gutiérrez y Muhs, G., Niemann, Y. F., González, C. G., & Harris, A. P. (2012). *Presumed incompetent: The intersections of race and class for women in academia*. Boulder, CO: Utah State University Press.

Hammond, L. H. (1916). *In the garden of delight*. New York, NY: Thomas Y. Crowell Company.

Hart, B., & Risley, T. (1995). *Meaningful differences in the everyday experience of young American children*. Baltimore, MD: Brookes Publishing.

Hauhart, R. C., & Grahe, J. E. (2015). *Designing and teaching undergraduate capstone courses*. San Francisco, CA: Jossey-Bass.

Hernstein, R. J., & Murray, C. (1994). *The bell curve: Intelligence and class structure in American life*. New York, NY: Free Press Paperbacks.

Hilliard, A. G., III. (1990). Back to Binet: The case against the use of IQ tests in the schools. *Contemporary Education*, 184–189.

Hixenbaugh, M. (2011, September 17). Teacher-student racial imbalance widest in Virginia Beach. *The Virginian-Pilot*. Retrieved from pilotonline.com/news/local/education/teacher-student-racial-imbalance-widest-in-va-beach/article_99d4e10f-7582-5625-b2c5-549983e5f028.html

Horne, J. A., & Ostberg, O. (1976). A self-assessment questionnaire to determine morningness-eveningness in human circadian rhythms. *International Journal of Chronobiology*, 4, 97–110.

Horsford, S. D. (2011). *Learning in a burning house: Educational inequality, ideology, and (dis)integration*. New York, NY: Teachers College Press.

Hughes, L. (1926, June 23). "The Negro artist and the racial mountain." *The Nation, 122*, 692–694.

Hunter, A., Laursen, S. L., & Seymour, E. (2007). Becoming a scientist: The role of undergraduate research in students' cognitive, personal, and professional development. *Science Education, 91*(1), 36–74.

Hyde, C. R. (1999). *Pay it forward*. New York, NY: Simon & Schuster.

Inzlicht, M., & Ben-Zeev, T. (2003). Do high-achieving female students underperform in private? The implications of threatening environments on intellectual processing. *Journal of Educational Psychology, 95*(4), 796–805.

It Gets Better Project. (2010–2016). It gets better project. Retrieved from /www.itgetsbetter.org/

Jaschik, S. (2015, August 24). Who gets credit? *Inside Higher Ed*. Retrieved from www.insidehighered.com/news/2015/08/24/research-reveals-significant-share-scholarly-papers-have-guest-or-ghost-authors

Jones, C. P. (2002). Confronting institutionalized racism. *Phylon, 50*(1), 7–22.

Jones, J. M. (1997). *Prejudice and racism* (2nd ed.). New York, NY: McGraw-Hill.

Jones, M. T., Barlow, A. E. L., & Villarejo, M. (2010). Importance of undergraduate research for minority persistence and achievement in biology. *Journal of Higher Education, 81*(1), 82–115.

Kahneman, D., & Tversky, A. (1979). Prospect theory: An analysis of decision under risk. *Econometrica, 47*(2), 263–291.

Kardash, C. M. (2000). Evaluation of an undergraduate research experience: Perceptions of undergraduate interns and their faculty mentors. *Journal of Educational Psychology, 92*(1), 191–201.

Kennedy, J. F. (1961, July 25). Proclamation 3422, American Education Week 1961. *The American Presidency Project.* Retrieved from www.presidency.ucsb.edu/ws/?pid=24146

Kinkel, D. H., & Henke, S. E. (2006). Impact of undergraduate research on academic performance, educational planning, and career development. *Journal of Natural Resources & Life Sciences Education, 35*(1), 194–201.

Kost-Smith, L. E., Pollock, S. J., & Finkelstein, N. D. (2010). Gender disparities in second-semester college physics: The incremental effects of a "smog of bias." *Physical Review Special Topics—Physics Education Research, 6.*

Kuh, G. D., & O'Donnell, K. (2013). *Ensuring quality and taking high-impact practices to scale.* Washington, D.C.: AAC&U.

Kuh, G. D., Kinzie, J., Schuh, J. H., Whitt, E. J., & Associates. (2010). *Student success in college: Creating conditions that matter* (2nd ed.). San Francisco, CA: Jossey-Bass.

Labov, W. (1982). Objectivity and commitment in linguistic science: The case of the Black English trial in Ann Arbor. *Language in Society 11*(2), 165–201.

Lambert, E. A., (2016). Unpacking the psychosocial effects of institutional racism. *College of William & Mary Undergraduate Honors Theses.* Paper 960.

Laursen, S., Hunter, A. B., Seymour, E., Thiry, H., & Melton, G. (2010). *Undergraduate research in the sciences: Engaging students in real science.* San Francisco, CA: Jossey-Bass.

Lepore, S. J., Revenson, T. A., Weinberger, S. L., Weston, P., Frisina, P. G., Robertson, R., . . . Cross, W. (2006). Effects of social stressors on cardiovascular reactivity in Black and White women. *Annals of Behavioral Medicine, 31*(2), 120–127.

Lippi-Green, R. (2012). *English with an accent: Language, ideology, and discrimination in the United States* (2nd ed.). New York, NY: Routledge.

Lopatto, D. (2007). Undergraduate research experiences support science career decisions and active learning. *CBE Life Sciences Education, 6*, 297–306.

Lupton, D. (2014). *'Feeling better connected': Academics' use of social media.* Canberra, Australia: News & Media Research Centre, University of Canberra.

Mabrouk, P. A., & Peters, K. (2000). Student perspectives on undergraduate research (UR) experiences in chemistry and biology. *CUR Quarterly, 21*(1), 25–33.

Malachowski, M. (1996). The mentoring role in undergraduate research projects. *Council on Undergraduate Research Quarterly, 1996* (December), 91–93, 105–106.

Mandela, N. (2003, July 16). Lighting your way to a better future [speech]. *Nelson Mandela Foundation*. Retrieved from www.mandela.gov.za/mandela_speeches/2003/030716_mindset.htm

Martens, A., Johns, M., Greenberg, J., & Schimel, J. (2006). Combating stereotype threat: The effect of self-affirmation on women's intellectual performance. *Journal of Experimental Social Psychology, 42*, 236–243.

Maslow, A. H. (1943). A theory of human motivation. *Psychological Review, 50*(4), 370–396.

McCraney, T. A. (2013). *Marcus; Or the secret of sweet*. New York, NY: Dramatists Play Service Inc.

McNair Scholars Program. (n.d.) *McNair Scholars Program*. Retrieved from mcnairscholars.com/

Medin, D. L., & Lee, C. D. (2012). Diversity makes better science. *Observer, 25*(5). Retrieved from www.psychologicalscience.org/index.php/publications/observer/2012/may-june-12/diversity-makes-better-science.html

Moss-Racusin, C. A., Dovidio, J. F., Brescoll, V. L., Graham, M. J., & Handelsman, J. (2012). Science faculty's subtle gender biases favor male students. *Proceedings of the National Academy of Sciences, 109*(41), 16474–16479.

National Academies of Sciences, Engineering, and Medicine. (2016a). *Ford Foundation fellowship programs*. Retrieved from sites.nationalacademies.org/pga/fordfellowships/index.htm

National Academies of Sciences, Engineering, and Medicine. (2016b). *Predoctoral fact sheet*. Retrieved from sites.nationalacademies.org/pga/fordfellowships/pga_047958

National Center for Faculty Development and Diversity. (2011). *NCFDD mentoring map*. Retrieved from dl.dropboxusercontent.com/u/72986838/Frequent%20Downloads/Mentoring%20Map.pdf

National Endowment for the Arts. (n.d.). *National Endowment for the Arts*. Retrieved from www.arts.gov/

National Endowment for the Humanities. (n.d.). *Documenting endangered languages*. Retrieved from www.neh.gov/grants/preservation/documenting-endangered-languages

National Endowment for the Humanities. (n.d.). *Humanities initiatives at Hispanic-Serving Institutions*. Retrieved from www.neh.gov/grants/education/humanities-initiatives-hispanic-serving-institutions

National Endowment for the Humanities. (n.d.). *Humanities initiatives at Historically Black Colleges and Universities*. Retrieved from www.neh.gov/grants/education/humanities-initiatives-historically-black-colleges-and-universities

National Endowment for the Humanities. (n.d.). *NEH impact reports*. Retrieved from www.neh.gov/news/impact-reports

National Endowment for the Humanities. (n.d.). *Public scholar program*. Retrieved from www.neh.gov/grants/research/public-scholar-program

National Endowment for the Humanities. (n.d.). *Sustaining cultural heritage col-*

lections. Retrieved from www.neh.gov/grants/preservation/sustaining-cultural-heritage-collections

National Institute on Minority Health and Health Disparities. (n.d.). "Goals." *Community-based participatory research program.* Retrieved from www.nimhd.nih.gov/programs/extramural/community-based-participatory.html#goals

National Science Foundation. (2016). *Proposal and award policies and procedures guide.* Retrieved from www.nsf.gov/pubs/policydocs/pappguide/nsf16001/gpg_print.pdf

National Science Foundation. (n.d.). *Research Experiences for Undergraduates.* Retrieved from www.nsf.gov/funding/pgm_summ.jsp?pims_id=5517&from=fund

The Ohio State University. (2016). *Summer Research Opportunities Program.* Retrieved from gradsch.osu.edu/outcomes-innovations/summer-research-opportunities-program

Painter, N. I. (2010). *The history of White people.* New York, NY: W. W. Norton & Company, Inc.

Paris, D., & Winn, M. (Eds.) (2012). *Humanizing research. Decolonizing qualitative inquiry with youth and communities.* Thousand Oaks, CA: Sage Publications.

Petrie, M. (2007, September 21). NIH seeks to foster interdisciplinary research. *Chronicle of Higher Education.* Retrieved from chronicle.com.article/NIH-Seeks-to-Foster/5261

Petrie, T. A., Hankes, D. M., & Denson, E. L. (2011). *A student athlete's guide to college success: Peak performance in class and life* (3rd ed.). Boston, MA: Wadsworth Cengage Learning.

Pollak, L. (2012). *Getting from college to career: Your essential guide to succeeding in the real world.* New York, NY: HarperCollins.

Posse Foundation. (2014). *The Posse Foundation, Inc.* Retrieved from www.posse-foundation.org/

Regents of the University of Michigan. (2016). *Undergraduate Research Opportunity Program.* Retrieved from lsa.umich.edu/urop

Rice, K. G., Richardson, C. M. E., & Tueller, S. (2013). The short form of the revised almost perfect scale. *Journal of Personality Assessment, 96*(3), 368–379.

Rockquemore, K. A. (2016, February 17). Can I mentor African-American faculty? *Inside Higher Ed.* Retrieved from www.insidehighered.com/advice/2016/02/17/advice-white-professor-about-mentoring-scholars-color-essay

Rockquemore, K. A., & Laszloffy, T. (2008). *The Black academic's guide to winning tenure.* Boulder, CO: Lynne Rienner.

Rothblum, E. D., Solomon, L. J., & Murakami, J. (1986). Affective, cognitive, and behavioral differences between high and low procrastinators. *Journal of Counseling Psychology, 33*(4), 387–394.

Russell, S. H., Hancock, M. P., & McCullough, J. (2007, April 27). Benefits of undergraduate research experiences. *Science, 316*(5824), 548–549.

Ryan, J. (2010). *Five miles away, a world apart: One city, two schools, and the story of educational opportunity in America*. New York, NY: Oxford University Press.

Saddler, T. N. (2010). Mentoring and African American undergraduates' perceptions of academic success. In T. L. Strayhorn & M. C. Terrell (Eds.), *The evolving challenges of Black college students: New insights for policy, practice, and research* (pp. 179–200). Sterling, VA: Stylus Publishing.

Saenz, D. S. (1994). Token status and problem-solving deficits: Detrimental effects of distinctiveness and performance monitoring. *Social Cognition, 12*, 61–74.

Schmidt, R. (2002). Radicalization and language policy: The case of the U.S.A. *Multilingua, 21*, 141–161.

Seal, D. (2015, April 19). Black Student Alliance calls for sweeping changes at UVA. *The Daily Progress*. Retrieved from www.dailyprogress.com/news/local/black-student-alliance-calls-for-sweeping-changes-at-uva/article_e9015b1a-e6ff-11e4-bac8-53e51cc93916.html

Seeman, J. I., & House, M. C. (2010). Influences on authorship issues: An evaluation of giving credit. *Accountability in Research, 17*, 146–169.

Sekaquaptewa, D., & Thompson, M. (2002). The differential effects of solo status on members of high- and low-status groups. *Personality and Social Psychology Bulletin, 28*(5), 694–707.

Silvia, P. (2007). *How to write a lot: A practical guide to productive academic writing*. Washington, DC: American Psychological Association.

Slaney, R. B., Mobley, M., Trippi, J., Ashby, J. S., & Johnson, D. (1996). The almost perfect scale revised. Retrieved from kennethwang.com/apsr/scales/APS-R_96.pdf

Smith, B. (2013). *Mentoring at-risk students through the hidden curriculum of higher education*. Lanham, MD: Lexington Books.

Steele, C. M. (1992, April). Race and the schooling of Black Americans. *The Atlantic Monthly*. Retrieved from www.theatlantic.com/past/docs/unbound/flashbks/blacked/steele.htm

Steele, C. M. (1999, August). Thin ice: Stereotype threat and Black college students. *The Atlantic Monthly*. Retrieved from www.theatlantic.com/magazine/archive/1999/08/thin-ice-stereotype-threat-and-black-college-students/304663/

Steele, C. M. (2010). *Whistling Vivaldi: How stereotypes affect us and what we can do*. New York, NY: W. W. Norton & Company.

Stokes, S. (2013, November 4). *The Black Bruins* [Spoken word]. Retrieved from www.youtube.com/watch?v=BEO3H5BOlFk

Sue, D. W. (2010). *Microaggressions in everyday life: Race, gender, and sexual orientation*. Hoboken, NJ: John Wiley & Sons.

Thomson, P., & Kamler, B. (2013). *Writing for peer reviewed journals: Strategies for getting published*. New York, NY: Routledge.

Title IX of the Education Amendments of 1972, 20 U.S.C. a§ 1681 et seq. (1972).

Triandis, H., Bontempo, R., & Villareal, M. (1988). Individualism and collectivism: Crosscultural perspectives on self-ingroup relationships. *Journal of Personality and Social Psychology, 54*(2), 323–338.

Trower, C., & Chait, R. (2002). Faculty diversity: Too little for too long. *Harvard Magazine*. Retrieved from harvardmagazine.com/2002/03/faculty-diversity.html

U.S. Department of Education Office for Civil Rights. (2016). How to file a discrimination complaint with the Office for Civil Rights. Retrieved from www2.ed.gov/about/offices/list/ocr/docs/howto.html

U.S. Department of Education, National Center for Education Statistics, Integrated Postsecondary Education Data System (IPEDS). (2013). *Percentage of total and STEM bachelor's degrees conferred by postsecondary institutions, by race/ethnicity and gender: 2012–13*. Retrieved from nces.ed.gov/programs/raceindicators/indicator_reg.asp

U.S. News & World Report L.P. (2016). *Best colleges*. Retrieved from colleges.usnews.rankingsandreviews.com/best-colleges

U.S. News & World Report L.P. (2016). *Best undergraduate teaching, national universities*. Retrieved from colleges.usnews.rankingsandreviews.com/best-colleges/rankings/national-universities/undergraduate-teaching

Van Blerkom, D. (2013). *Orientation to college learning* (7th ed.). Boston, MA: Wadsworth Cengage Learning.

Verrell, P. A., & McCabe, N. R. (2015). In their own words: Using self-assessments of college readiness to develop strategies for self-regulated learning. *College Teaching, 63*(4), 162–170.

Vohs, K. D., Ciarocco, N. J., & Baumeister, R. F. (2005). Self-regulation and self-presentation: Regulatory resource depletion impairs impression management and effortful self-presentation depletes regulatory resources. *Journal of Personality & Social Psychology, 88*(4), 632–657.

Vye, C., Scholljegerdes, K., & Welch, I. D. (2007). *Under pressure and overwhelmed: Coping with anxiety in college*. Westport, CT: Praeger.

Wachtel, T. (2013). *Defining restorative*. Bethlehem, PA: International Institute for Restorative Practices.

Walsh, J. P. & Jabbehdari, S. (2015, August 22). In the club: Authorship norms and collaboration structures in science. Paper presented at the American Sociological Association Annual Meeting, Chicago, IL.

Walton, G. M., & Cohen, G. L. (2011, March 18). A brief social-belonging intervention improves academic and health outcomes of minority students. *Science 331*(6023), 1447–1451.

Watson, A., Bristol, T., White, T., & Vilson, J. L. (2015, July 14). Recruiting and retaining educators of color. [Blog]. Retrieved from www.shankerinstitute.org/blog/recruiting-and-retaining-educators-color

Wisker, G. (2005). *The good supervisor: Supervising postgraduate and undergraduate research for doctoral theses and dissertations*. New York, NY: Palgrave Macmillan.

Wolfe, G. C. (1988). *The colored museum*. New York, NY: Grove Press.
Ziersch, A. M., Gallaher, G., Baum, F., & Bentley, M. (2011). Responding to racism: Insights on how racism can damage health from an urban study of Australian aboriginal people. *Social Science & Medicine, 73*(7), 1045–1053.
Zuberi, T. (2000). *Thicker than blood: How racial statistics lie*. Minneapolis, MN: University of Minnesota Press.

Index

About the Authors

Anne H. Charity Hudley is Class of 1952 Term Associate Professor of English, Linguistics, Africana Studies, and Education at the College of William & Mary in Williamsburg, VA. She also directs the William & Mary Scholars Program and co-directs the William & Mary Scholars Undergraduate Research Experience (WMSURE). She is on the executive committee of the Linguistic Society of America. In July 2017, she will become the North Hall Endowed Chair in the Linguistics of African-America and Director of Undergraduate Research at the University of California at Santa Barbara. Her two previous books and 25 publications address the relationship between English language variation and K–16 educational practices and policies. She has earned numerous awards for teaching and is associate editor of *Language* with specific responsibilities for articles concerning the teaching of linguistics. She has worked with K–12 educators through lectures and workshops sponsored by the American Federation of Teachers and by public and independent schools throughout the country.

Dr. Charity Hudley earned a BA and an MA in Linguistics from Harvard University in 1998. She earned a PhD in Linguistics from the University of Pennsylvania in 2005. She was awarded a Ford Pre-Dissertation Fellowship in 2003. From 2003–2005, she was the Thurgood Marshall Dissertation Fellow at Dartmouth College. She received a National Science Foundation Minority Postdoctoral Fellowship in Fall 2005, a National Science Foundation Minority Research Starter Grant in 2009, and an NSF Collaborative grant in 2013 to study language and culture in STEM contexts. Charity Hudley was also the principal investigator of grants from the Jessie Ball duPont Fund and Bank of America to improve the WMSURE program.

Cheryl L. Dickter is the Wilson P. and Martha Claiborne Stephens Term Distinguished Associate Professor of Psychology at the College of William & Mary. She is a faculty affiliate of the Neuroscience Program and the Gender, Sexuality, and Women's Studies Program. She co-directs the William & Mary Scholars Undergraduate Research Experience (WMSURE). She received her PhD in Social Psychology from the University of North Carolina in 2006. Her research uses a social cognitive neuroscience approach to examine how individuals perceive members of different social groups, and how these perceptions differ based on contextual information such as stereotypes. Dr. Dickter also examines how the cognitive processes involved in the processing of drug-related stimuli are affected by exposure, craving, and motivation. Her work has been funded by the National Science Foundation and the National Institutes of Health. She is an Associate Editor for *Social Cognitive and Affective Neuroscience,* is on the editorial board for *Social Cognition,* and is an NIH grant reviewer for the Risk, Prevention, and Health Behavior section of the Center for Scientific Review. Dr. Dickter has received three college-wide teaching awards as well as a national award for her work as the faculty advisor of the Psychology Department's honor society.

Hannah A. Franz is a PhD student in Educational Policy, Planning, and Leadership at the College of William & Mary. She is a graduate assistant for the William & Mary Scholars Undergraduate Research Experience (WMSURE) and teaches in the Africana Studies program. Her research addresses equity in the transition from K–12 to college research and writing. From 2010–2013, she taught middle school reading in Norfolk (VA) Public Schools. Hannah earned an MSEd in Reading/Writing/Literacy from the University of Pennsylvania in 2010, an MA in Sociolinguistics from North Carolina State University in 2009, and a BA in Linguistics from the College of William & Mary in 2007.